W0042515

1992-93

**Statistics and Analyses
of the World's
Minerals Industry**

Minerals Handbook 1992-93

Statistics and Analyses of the World's Minerals Industry

Phillip Crowson

M
stockton
press

© Phillip Crowson, 1992.

Softcover reprint of the hardcover 1st edition 1992

All rights reserved. No part of this publication may be reproduced or transmitted, in any form or by any means, without permission.

Published in the United States and Canada by
STOCKTON PRESS, 1992
257 Park Avenue South, New York, N.Y. 10010

The Library of Congress has catalogued this serial publication as follows:

Minerals handbook (New York, N.Y.)
      Minerals handbook/compiled by Phillip Crowson. – 1982-83–
      New York: Van Nostrand Reinhold, c1982-

      (v. :ill. ; 23cm.)

      Biennial.
      Issued by: Gulf Pub. Co., 1984-85-
      Also published in the U.K. by Macmillan.
      ISSN 0265-3923 = Minerals handbook.
      1. Mines and mineral resources - Handbooks, manuals, etc. 1.
      Crowson, Phillip. II. Gulf Publishing Company.

      TN151.M49   333.8'5–dc19   85-645605
                                           AACR 2   MARC-S
      Library of Congress   8512

      ISBN 978-1-56159-054-4

First published in the United Kingdom by
MACMILLAN PUBLISHERS LTD, 1992
Distributed by Globe Book Services Ltd
Brunel Road, Houndmills,
Basingstoke, Hampshire RG21 2XS

British Library Cataloguing in Publication Data
Minerals handbook. – 1992-93
      1. Mineral industries & trades -
      Statistics - Serials
      338.2'0212

      ISBN 978-1-349-12566-1   ISBN 978-1-349-12564-7
      DOI 10.1007/978-1-349-12564-7
      ISSN 0265-3923

CONTENTS

CONTENTS

Page

INTRODUCTION

This is the Sixth Edition of this Handbook in its present form. It originally developed from a study on Non-Fuel Minerals and British Foreign Policy prepared in 1978 for the Royal Institute of International Affairs. Each Edition has been expanded and refined, to the point where this one contains data for forty-eight commodity groups.

All the data have again been updated and revised to include annual figures for 1989-90. The Handbook is not intended as a substitute for the many excellent statistical publications listed in the Appendix, from which its data are derived, but merely as an introductory guide mainly for the non-specialist. It draws together in a convenient form information that is scattered over a very wide range of primary and secondary sources. The numerous statistical caveats, qualifications and footnotes accompanying the original sources of the statistics have been omitted. The tables would otherwise have been swamped in a sea of footnotes.

One purpose of the handbook is to contain sufficient basic data on all aspects of the minerals and metals included to allow informed debate on mineral policies. Another is to give reasonably comprehensive introductions to each material covered.

The earlier versions emphasised that "the real world is invariably far more complex than simple tables might suggest, and the tendency to latch on to published statistics as if they were unshakeable truths should be avoided". To illustrate this point, "statistics on reserves of any mineral involve extensive inference from incomplete data and extensive judgement, not just about the technical characteristics of ore deposits, but also about their economics". Even many estimates of production and trade carry wide margins of error.

The book's layout is straightforward; the introductory summary tables are followed by separate sections on each of the forty-eight minerals. The summary tables mainly bring together data contained in the detailed sections, but with the addition of a table showing the approximate turnover and growth rates of the world's mineral industries, and another showing the historic growth in reserves of several minerals. The other tables summarise aspects of mineral industries that influence public policy. Thus Table 2 shows two measures of the adequacy of reserves and Table 5 contains estimates of import dependence in the main areas. The table which showed South Africa's role in the world mineral industries has been dropped from this edition, in the light of political developments in that country. The basic data are, however, still available in the individual sections on each commodity.

The individual sections in each mineral follow a broadly common format. The varying units of the sources have been converted into metric equivalents for nearly all the metals, although imperial and apothecaries' measures are the conventional measures in many cases. Prices are, however, usually quoted in their original units. The main sources of the statistics used are summarised in a section at the end of the book. The available data, and the specific characteristics of each mineral, explain any variations from the standard pattern, which is as follows.

The tables on reserves and production still divide the world into three conventional groupings. With the political changes of recent years, and particularly with the break up of the USSR, the disbandment of Comecon, and the unification of Germany, the subdivision is no longer very useful. It was, however, relevant for the period covered by the tables and is retained. Separate data continue to be shown for West and East Germany. The three broad groups of countries used are **Developed, Developing**, and **Centrally Planned**. The Developed Countries are arbitrarily defined as all OECD Members plus South Africa, the Centrally Planned are the former USSR, the former Comecon Countries, the People's Republic of China, North Korea, North Vietnam, Albania, Mongolia and Cuba. Developing Countries are all those not included in the other two categories.

World Reserves and Reserve Bases

The statistics are taken mainly from publications of the US Bureau of Mines and private communications with Bureau staff. The figures refer mainly to 1990-91, but with figures for earlier years in some tables.

Reserves are defined as:

Recoverable materials in the reserve base that can be economically extracted or produced at the time of determination.

The *Reserve base* is more broadly described as:

In-place demonstrated (measured plus indicated) resource from which reserves are estimated, and including those resources that are currently economic (reserves), marginally economic (marginal reserves), and some of those that are currently subeconomic (subeconomic reserves).

Where practicable details are given on the narrower definition, but in one or two instances both measures are included. Usually a footnote gives a broader estimate of total resources, which include mineral deposits that are not presently economic. As noted earlier, changes in the basic technical and economic assumptions can dramatically alter estimates of reserves; there is no objective measure. Summary Table 3 demonstrates that estimated reserves in most instances rose over time at least as fast as production.

World Production

Again there are separate figures for each producing country, subdivided into the three broad groupings. The sources are broadly similar to those for reserves. Separate figures are given for 1989 and 1990. For some commodities figures are given for the mined product and for its main derivatives. Thus there are separate tables for bauxite, alumina and aluminium.

Productive Capacity

For the major producing countries estimates of productive capacity are included for 1990. They are mainly taken from recently published data from the US Bureau of Mines.

Secondary Production

Where recycled material is important, and the statistics are available, a separate table shows average supplies in the main areas in 1989 and 1990.

The Adequacy of Reserves

Two estimates are given, based mainly on the earlier tables. The first is the static life of identified reserves, assuming that production continues at the 1989-90 level. In most instances production is gowing however, and reserves are also tending to rise as new deposits are discovered, and existing mines extend their knowledge. The second figure, to a certain extent, allows for these changes. It shows the ratio of the identified reserve base (which is greater than reserves) to cumulative demand between 1991 and 2010 (based mainly on historic growth rates projected forward). Whilst this dynamic ratio is more meaningful than the static reserve/production ratio, it should, nonetheless, be treated with considerable caution. A broader ratio of resources to cumulative demand would normally be much greater. As prices rise or costs fall more deposits will move from resources into reserves thereby sustaining supplies for longer than the printed ratios suggest.

Consumption

Consumption in 1989 and 1990 is given for the main consuming areas, and historic annual average compound growth rates, in all cases during the 1970s and 1980s and wherever possible for the 1960s as well. The underlying data have been derived from a wide variety of sources.

End Use Patterns

Data are given for the United States in 1990 and in a few cases for other countries, although such details are not as readily available. Although consumption patterns differ in detail for the rest of the world the US figures are reasonably representative of the main outlets for each material. In some sections, and particularly where there are important non-metallic as well as metallic uses, separate figures are given for the mineral and for its immediate products.

Value of Contained Metal in Annual Production

Total world production is multiplied by the average prices of 1991 to indicate the product's relative importance as an article of commerce. These values, however approximate, are better guides than relative tonnages alone.

Substitutes and Technical Possibilities

Based largely on the reports of the US Bureau of Mines, these two sections pinpoint how supply and demand may change.

Prices

A description of the pricing methods employed, with a table showing annual average prices between 1986 and 1991. For some commodities only one price is given, whilst in other cases there are several. The prices are taken from various trade publications. Each table is supplemented by a chart which shows the trend in a representative price since 1979. The charts plot index numbers (1991 = 100) of money prices and of prices in 'real' December 1991 terms. In order to obtain the latter the money prices are deflated by a relevant wholesale price index. This is the United States' index of producer prices of industrial commodities where prices are given in dollars, as they are in most instances.

Marketing Arrangements

A brief description of the structure of the market, and of any relevant international organisations.

Supply and Demand by Main Market Area

Domestic production, trade and consumption in the United Kingdom, European Community, Japan and United States. Domestic production is divided into the main stages where relevant. The source of net imports (i.e. imports from third countries) are shown, and also dependence on external supplies. The data are mainly for 1989 and 1990. The geographical sources of net imports are given as percentages of the total separately for all four groupings. Shares of world consumption and the historic growth of consumption are also included. For most metals the main additional sources to those used for the earlier tables are the relevant trade statistics.

The geographical coverage of the European Community has changed frequently. In this Handbook it is the Europe of the Twelve which was the coverage for most of the statistics included.

ACKNOWLEDGEMENTS

The compiler gratefully acknowledges indebtedness to the statistical publications of the US Bureau of Mines, The World Bureau of Metal Statistics, Metallgesellschaft, and the British Geological Survey in particular. Many other sources, listed at the end of the report, have also been used. The price data are derived mainly from the Metal Bulletin, Metals Week, Industrial Minerals and the Engineering and Mining Journal.   Any mistaken interpretations, errors or omissions, are the compiler's sole responsibility. A considerable proportion of the data was put together by Liz Power and Celia Prentice, without whom the book would never have been produced. Thanks are also due to the staff of RTZ Japan and in particular to Ken Suzuki and Hiroko Tsuruoka. Finally Elsie Arjune has undertaken the arduous task of typing the data.

n.a	Not available
c.	approximately
..	Under 1

Independent rounding means that percentages may not add up to 100 throughout the publication.

TABLE 1

PRODUCTION AND RESERVES BY MAIN GEO-POLITICAL GROUPING

	% Share in World Reserves			Primary Production 1990		
	Developed	Developing	Centrally Planned	Developed	Developing	Centrally Planned
Bauxite	25	71	4	43	46	11
Alumina	-	-	-	60	23	17
Aluminium	-	-	-	63	18	19
Antimony	16	24	59	22	24	54
Arsenic (a)	n.a	n.a	n.a	49	34	16
Asbestos	45	9	46	23	12	64
Barytes	27	43	30	27	29	44
Beryllium	12	71	16	54	11	35
Bismuth	49	28	23	27	42	31
Boron	69	8	23	92	5	3
Cadmium (a)	59	28	13	64	14	22
Chromium	73	17	10	46	19	35
Cobalt	4	61	35	11	77	12
Copper	27	57	16	37	43	20
Fluorspar	25	12	63	21	23	56
Gold	67	18	5	65	17	17
Indium	60	21	19	83	3	14
Industrial Diamonds	58	37	5	37	44	18
Iron Ore	38	20	42	32	33	35
Kaolin	59	36	5	67	14	19
Lead	58	14	28	52	19	29
Lithium(b)	41	59	n.a	63	23	14
Magnesite	13	12	75	38	9	53
Manganese	50	12	38	29	32	39
Mercury	75	8	17	20	18	62
Molybdenum	57	26	17	68	19	13
Nickel	24	22	54	38	28	34
Niobium	3	81	16	18	82	n.a
Phosphate	34	52	14	22	73	5
Platinum Group	89	..	11	56	..	44
Potash	54	2	44	49	8	43
Rare Earths	23	25	52	58	11	31
Rhenium	17	59	24	53	23	24
Selenium	31	59	10	85	5	10
Silicon	n.a	n.a	n.a	59	16	25
Silver	44	36	20	41	41	18
Sulphur	31	31	38	52	14	34
Talc(b)	72	28	n.a	50	29	21
Tantalum	30	53	17	n.a	n.a	n.a
Tellurium (a)	28	62	10	n.a	n.a	n.a
Tin	8	61	31	8	67	25

	% Share in World Reserves			Primary Production 1990		
	Developed	Developing	Centrally Planned	Developed	Developing	Centrally Planned
Titanium:						
Ilmenite	63	20	17)	78	15	7
Rutile	11	86	3)			
Tungsten	29	11	60	11	9	80
Uranium	60	21	19	45	15	40
Vanadium	24	n.a	76	55	-	45
Vermiculite	n.a	n.a	n.a	79	10	11
Zinc	61	24	15	53	21	26
Zirconium	83	8	9	82	7	11

(a) Production at refineries.
(b) Western World only for reserves.

TABLE 2

THE 'ADEQUACY' OF RESERVES

	Static Reserve Life (years)	Ratio of Identified Reserve Base to Cumulative Primary Demand 1991-2010
Bauxite	220	14
Antimony	78	3.9
Arsenic	21	1.1
Asbestos	27	2.1
Barytes	30	4.4
Beryllium	very large	very large
Bismuth	28	3.6
Boron	295	23
Cadmium	27	2.1
Chromium	105	23
Cobalt (land only)	90	18
Copper (land only)	36	2.2
Fluorspar	52	4.4
Gallium	very large	very large
Germanium	large	large
Gold	22	0.7
Indium	17	0.6
Industrial Diamonds	18	0.7 (exc.synthetic)
Iron Ore	119	7.9
Kaolin	very large	33
Lead	20	1.1
Lithium	very large	40
Magnesium	very large	over 40 (exc. brines)
Manganese (land only)	95	23.4
Mercury	25	2.4
Molybdenum	50	6.1
Nickel (land only)	55	4.8
Niobium	over 300	1.1
Phosphate	very large	9
Platinum Group	197	19.5
Potash	300	30
Rare Earths	very large	50
Rhenium	88	12.4
Selenium	41	3.9
Silicon	extremely large	very large
Silver	18.5	2.2
Sulphur	24	2.9
Talc	46	6
Tantalum	75	1.4
Tellurium	102	7.5
Tin	28	1.8
Titanium	70	7.5
Tungsten	55	5.7
Uranium	58	3.7
Vanadium	135	23
Vermiculite	81	13.6
Zinc	21	2.1
Zirconium	55	3.8

TABLE 3

THE GROWTH OF WORLD RESERVES OF SELECTED PRODUCTS

This table shows how estimates of world reserves of four major base metals increased over a forty year period relative to the rate of growth of world mine production. Figures for these metals are more readily available than for many others, but in most respects the pattern shown is typical; estimated reserves grew at least as fast as production until the 1980s.

(million tonnes contained metal near the end of the relevant decade)

	Copper	Lead	Zinc	Aluminium (a)
1940s	91	91	54 to 44	1605
1950s	124	45 to 54	77 to 86	24164
1960s	280	86	106	11600
1970s	543	157	240	22700
1980s(b)	566	120	295	23200
% p.a. growth				
1950s-1970s	7.5	5 to 5.75	9.75	
1980s	0.4	-2.7	2.1	0.2
% p.a. growth of mine production				
1950s-1970s	3.75	1.75	2.75	7.0
1980s	1.4	-0.6	1.0	1.7

(a) Gross weight of bauxite.
(b) Reserve base in 1989.

In the first half of the 1980s a decline in prices relative to costs led to reductions in reserves, or hitherto economic ore bodies became uneconomic. The position was not fully reversed in the second half of the decade. Hence production tended to rise more rapidly than reserves in the 1980s, but not by enough to invalidate the longer term trends.

TABLE 4

VALUE OF CONTAINED METAL IN ANNUAL PRODUCTION
million US $

Aluminium	24000	
Antimony	90	
Arsenic	25	(arsenic trioxide)
Asbestos	1500	
Barytes	500	
Beryllium	220	
Bismuth	26	
Boron	900	
Cadmium	90	(refined metal)
Chromium	700	
Cobalt	950	(refined metal)
Copper	25000	(refined metal)
Fluorspar	710	
Gallium	21	
Germanium	50	
Gold	24000	
Indium	39	
Industrial Diamonds	464	
Iron Ore	18000	
Kaolin	2500	
Lead	3200	(refined metal)
Lithium	210	
Magnesite	1500	(exc. brines)
Magnesium metal	1130	
Manganese	3400	(metal content)
Mercury	19	
Molybdenum	580	
Nickel	7000	(refined metal)
Niobium	120	(Western World)
Phosphate	5600	
Platinum Group	3090	
Potash	2600	
Rare Earths	105	(concentrates)
Rhenium	39	
Selenium	43	(refined metal)
Silicon	5500	
Silver	2000	
Sulphur	5100	
Talc	1000	
Tantalum	45	
Tellurium	15	(refined metal Western World)
Tin	1200	(refined metal)
Titanium	1000	
Tungsten	244	
Uranium	2200	
Vanadium	400	
Vermiculite	50	
Zinc	10800	(refined metal)
Zirconium	272	

This table is merely designed to give a rough indication of the relative importance of the different minerals, on the basis of estimated turnover. These are based on average 1991 prices and 1990 world production levels.

TABLE 5
IMPORT DEPENDENCE 1990 IN PERCENTAGES
Imports as a percentage of domestic consumption plus exports

	United Kingdom	European Community	Japan	United States
Aluminium (inc. bauxite and alumina)	88	81	80	65
Antimony (a)	100	100	100	100
Arsenic	100	25	51	100
Asbestos	100	72	100	59
Barytes	76	44	100	72
Beryllium	100	100	100	11
Bismuth	100	n.a	38	98
Boron	100	100	100	-
Cadmium (refined)	61	28	43	46
Chromium	100	96	99	100
Cobalt (a)	100	100	100	100
Copper	96	71	95	20
Fluorspar	13	34	100	85
Gallium	n.a.	n.a	n.a	n.a
Germanium (refined)	100	n.a	18	n.a
Indium	n.a.	n.a	60	50
Iron Ore	82	94	100	22
Kaolin	-	14	90	-
Lead	57	39	65	7
Lithium	100	100	100	15
Magnesium Metal	65	58	35	13
Manganese	100	99	100	100
Mercury	100(a)	n.a	14	1
Molybdenum	100	100	100	-
Nickel	100	92	100	100
Niobium	100	100	100	100
Phosphate	100	100	100	-
Platinum Group (a)	100	100	100	100
Potash	34	27	100	70
Rare Earths	100	100	100	-
Rhenium	100	100	100	-
Selenium	100	70	7	38
Silicon(inc. ferrosilicon)	100	51	90	43
Silver	100	63	88	30
Sulphur	72	19	10	18
Talc	85	37	92	6
Tantalum	100	100	100	100
Tellurium	100	47	10	under 32
Tin	44	78	97	83
Titanium	100	100	100	100
Tungsten	99(a)	89(a)	91(a)	87
Uranium(ignoring stocks)	100	80	100	n.a
Vanadium	100	100	100	61
Vermiculite	100	100	n.a	18
Zinc	82	59	73	42
Zirconium	100	100	100	21

(a) Before allowing for secondary recovery.
In calculating these ratios no allowance has generally been made for changes in stocks.

TABLE 6

THE HISTORIC GROWTH OF TOTAL MINE PRODUCTION
% p.a. average compound growth rates

	1970s	1980s
Aluminium (bauxite)	5.4	1.1
Antimony	0.7	-0.1
Arsenic	-4.5	3.7
Asbestos	3.4	-1.7
Barytes	5.7	-2.9
Beryllium	c. -4.3	-1.0
Bismuth	c. 2.0	0.6
Boron	4.8	1.0
Cadmium	2.0	0.8
Chromium	5.9	2.1
Cobalt	3.6	-1.5
Copper	3.0	1.5
Fluorspar	2.3	-0.9
Gallium	20.0	11.2
Germanium	2.9	-3.4
Gold	-2.0	5.6
Indium	-1.5	9.4
Industrial Diamonds (including synthetics)	6.8	8.3
Iron Ore	2.3	0.7
Kaolin	4.9	2.4
Lead	0.8	0.4
Lithium	5.3	approx.1.6
Magnesium (all forms)	0.2	-1.0
Manganese	1.3	-0.4
Mercury	-4.0	1.2
Molybdenum	3.4	0.4
Nickel	3.8	1.6
Niobium	4.6	-2.8
Phosphate	4.9	1.2
Platinum Group	6.8	2.1
Potash	4.2	0.2
Rare Earths	3.8	5.6
Rhenium	11.1	5.1
Selenium	2.9	0.8
Silicon	5.8	1.1
Silver	1.5	2.9
Sulphur	3.8	0.5
Talc	4.6	0.9
Tantalum	0.3	-0.6
Tellurium	3.0	-0.9
Tin	1.1	-0.9
Titanium	1.5	1.3
Tungsten	3.3	-2.7
Uranium	8.0	-3.9
Vanadium	8.0	-0.8
Vermiculite	2.7	0.9
Zinc	1.2	1.4
Zirconium	3.2 (exc. USA)	1.9

TABLE 7
COMPARATIVE GROWTH RATES OF CONSUMPTION IN THE 1970s
% p.a. average compound rates 1969-70 to 1979-80 in most cases

	United Kingdom	European Community	Japan	United States
Aluminium (inc. secondary)	-1.6	4.1	7.3	3.2
Antimony (primary)	-7.5	n.a	-6.9	-3.2
Arsenic	n.a	n.a	n.a	-2.4
Asbestos	-2.8	1.1	1.9	-3.9
Barytes	6.8	-0.6	0.4	8.9
Beryllium	n.a	n.a	n.a	-1.7
Bismuth	n.a	n.a	n.a	-0.6
Boron	n.a	n.a	1.2	3.6
Cadmium	-0.2	1.4	-3.8	-2.1
Chromium	-4.8	6.5	4.1	0.5
Cobalt	-0.5	0.2	0.7	-
Copper	-1.9	1.3	4.9	..
Fluorspar	0.2	-0.4	-0.5	-2.5
Gallium	n.a	n.a	n.a	21.4
Germanium	n.a.	n.a	2.1	4.0
Gold (industrial uses)	0.2	-1.0	2.2	-2.6
Indium	n.a	n.a	n.a	1.7
Industrial Diamonds (inc. synthetics)	n.a	n.a	13.9	7.4
Iron Ore	-2.2	-0.1	7.0	-1.9
Kaolin	n.a	n.a	n.a	2.1
Lead	-0.6	0.2	3.4	0.6
Lithium	n.a	n.a	11.7	5.2
Magnesium Metal	-0.5	-2.0	8.5	2.8
Manganese Ore	-2.7	-0.2	1.5	-6.0
Manganese Ferro	-4.7	0.5	2.1	-1.6
Mercury	9.7(a)	n.a	-11.5	-1.4
Molybdenum	-3.5	2.3	4.4	3.1
Nickel	-0.3	3.4	4.3	1.8
Niobium	-5.0	5 to 8	12.1(b)	4.1
Phosphate	0.6	1.7	0.8	4.3
Platinum Group	n.a	n.a	10.1	6.3
Potash	-0.7	1.5	1.0	4.5
Rare Earths	n.a	n.a	n.a	5.2
Rhenium	n.a	n.a	n.a	7.2
Selenium	2.6	n.a	1.0	-5.0
Silicon	-2.0	n.a	5.8	2.4
Silver (industrial uses)	-0.7	-1.9	3.9	0.8
Sulphur	-0.7	0.7	-1.4	3.4
Talc	n.a	n.a	n.a	2.1
Tantalum	n.a	n.a	13.6(c)	1.8
Tellurium	n.a	n.a	n.a	1.2
Tin	-4.9	-1.7	1.9	-1.6
Titanium	-1.3	-"	4.3	1.5
Tungsten	-8.3	-6.0	4.9	1.6
Uranium (civil usage)	n.a	14.3	25.3	10.6
Vanadium	-3.0	n.a	8.7	-0.3
Vermiculite	n.a	n.a	n.a	1.9
Zinc	-3.0	0.8	2.2	-2.4
Zirconium	-1.7	4.2	8.4	0.2

(a) Primary only. (b) Ferroniobium only. (c) Powder only.

TABLE 8
COMPARATIVE GROWTH RATES OF CONSUMPTION IN THE 1980s
%p.a. average compound growth rates 1980 to 1990

	United Kingdom	European Community	Japan	United States
Aluminium (inc. secondary)	0.4	4.0	4.2	1.0
Antimony (primary)	-2.1	n.a	n.a	1.9
Arsenic	n.a	n.a	n.a	5.3
Asbestos	-16.6	-11.6	0.4	-20.6
Barytes	4.7	-	2.4	-7.1
Beryllium	n.a	n.a	n.a	4.6
Bismuth	n.a	n.a	n.a	1.8
Boron	n.a	n.a	n.a	-0.8
Cadmium	-3.3	0.3	8.8	-1.3
Chromium	n.a	0.7	2.9	-4.6
Cobalt	n.a	n.a	2.7	0.7
Copper	-2.5	1.2	3.2	1.4
Fluorspar	-1.2	-0.6	3.4	-4.4
Gallium	n.a	n.a	16.6	1.1
Germanium	n.a	n.a	-5.2	1.9
Gold (industrial uses)	6.0	8.8	10.9	3.8
Indium	n.a	n.a	22	4
Industrial Diamonds (inc. synthetics)	n.a	n.a	5.9	7.2
Iron Ore	1.3	-0.2	-1.2	-3.7
Kaolin	n.a	n.a	n.a	4.7
Lead	0.2	0.6	0.6	4.7
Lithium	n.a.	n.a	.0	-2.2
Magnesium metal	-5.9	0.1	3.2	1.0
Manganese ore	0.3	-2.4	1.6	-7.4
Manganese ferro	-0.1	0.6	-5.9	-6.1
Mercury	n.a	n.a	-6.3	-9.7
Molybdenum	-0.2	1.0	3.2	-2.4
Nickel	3.6	3.5	2.9	-1.1
Niobium	-	-0.3	3.6	-0.3
Phosphate	n.a	-3.5	-3.3	0.3
Platinum Group	n.a	n.a	8.0	1.7
Potash	3.0	-0.1	0.4	-1.9
Rare Earths	n.a	n.a	n.a	4.8
Rhenium	n.a	n.a	n.a	8.8
Selenium	5.7	2.5 to 3	3.4	3.8
Silicon(all forms)	-1.4	0.4	2.8	1.3
Silver (industrial uses)	-0.1	-0.3	4.5	1.7
Sulphur	-6.0	-1.1	1.2	-0.6
Talc	n.a	n.a	n.a	1.0
Tantalum	n.a	n.a	n.a	-3.2
Tellurium	n.a	n.a	n.a	n.a
Tin (inc. secondary)	0.5	0.5	0.9	-1.7
Titanium (pigments)	0.6	1.9	2.8	2.5
Tungsten	n.a	n.a	6.7	-1.8
Uranium	n.a	4.5	3.8	7.6
Vanadium	n.a	n.a	n.a	-6.1
Vermiculite	n.a	n.a	n.a	-2.4
Zinc	0.4	1.0	0.8	1.3
Zirconium	-6	0.2	-3.4	-2.5

ALUMINIUM/BAUXITE/ALUMINA

WORLD RESERVES OF BAUXITE
(million tonnes and % of total)

Developed			Developing			Centrally Planned		
Australia	4440	(20.5)	Brazil	2800	(13.0)	China	150	(0.7)
Greece	600	(2.8)	Cameroon	680	(3.2)	Hungary	300	(1.4)
USA	38	(0.2)	Ghana	450	(2.1)	Romania	50	(0.2)
Yugoslavia	350	(1.6)	Guinea	5600	(25.9)	USSR	300	(1.4)
Other (inc.			Guyana	700	(3.2)			
France)	40	(0.2)	India	1000	(4.5)			
			Indonesia	750	(3.5)			
			Jamaica	2000	(9.3)			
			Sierra Leone	140	(0.6)			
			Surinam	575	(2.6)			
			Venezuela	320	(1.5)			
			Others	304	(1.4)			
Totals	5468	(25.3)		15319	(71.0)		800	(3.7)
Grand Total		21587						

The bauxite **reserve base** is estimated at 25,000 million tonnes, and total world resources are estimated at 55 to 75,000 million tonnes.

Based on existing recovery techniques, the recoverable aluminium content of the world's bauxite reserves is 1 billion tonnes for developed countries, and 3 billion for the developing. Including centrally planned economies the total recoverable aluminium content of world reserves is 4,250 million tonnes.

Total world resources of bauxite (reserves plus sub-economic and undiscovered deposits) are calculated at 8,000 million tonnes of recoverable aluminium, on the basis of present recovery techniques. (The USSR also produces aluminium from alunite and nepheline syenite so that the table under-estimates the USSR's available deposits of aluminium containing minerals.)

ALUMINIUM/BAUXITE/ALUMINA

BAUXITE: WORLD MINE PRODUCTION, 1989-1990
('000 tonnes and % of total 1990)

Developed	1989	1990	% 1990
Australia	38583	41391	(36.7)
France	720	490	(0.4)
Greece	2602	2504	(2.2)
Turkey	534	779	(0.7)
USA	670	495	(0.5)
Yugoslavia	3252	2951	(2.6)
Others	12	-	(-)
Totals	**46373**	**48610**	**(43.1)**

Developing	1989	1990	% 1990
Brazil	7894	9876	(8.8)
Dominican Rep.	165	85	(0.1)
Ghana	348	381	(0.3)
Guinea	17547	17524	(15.5)
Guyana	1340	1424	(1.3)
India	4345	4340	(3.8)
Indonesia	862	1206	(1.1)
Jamaica	9395	10937	(9.7)
Malaysia	355	398	(0.3)
Sierra Leone	1548	1445	(1.3)
Surinam	3457	3267	(2.9)
Venezuela	702	771	(0.7)
Others	108	110	0.1
Totals	**48066**	**51764**	**(45.9)**

Centrally Planned	1989	1990	% 1990
China	4800	4200	(3.7)
Hungary	2643	2559	(2.3)
Romania	313	204	(0.2)
USSR	5750	5350	(4.8)
Albania	35	26	
Totals	**13541**	**12339**	**(11.0)**

Grand Totals 1989 - 107980
1990 - 112713

Note:The USSR's production of nepheline syenite and alunite was equivalent to roughly 1.1 million tonnes of bauxite in 1990.

ALUMINA: WORLD REFINERY PRODUCTION, 1989-90
('000 tonnes and % of total 1990)

Developed	1989	1990	% 1990
Australia	10823	11231	(27.0)
Canada	1048	1087	(2.6)
France	624	606	(1.5)
W Germany	1174	1165	(2.8)
Greece	533	585	(1.4)
Ireland	891	927	(2.2)
Italy	722	752	(1.8)
Japan	863	890	(2.1)
Spain	949	1002	(2.4)
Turkey	201	177	(0.4)
UK	116	131	(0.3)
USA	5180	5430	(13.0)
Yugoslavia	1240	1086	(2.6)
Totals	**24364**	**25069**	**(60.1)**

Developing	1989	1990	% 1990
Brazil	1624	1653	(4.0)
Guinea	627	642	(1.5)
India	1419	1334	(3.2)
Jamaica	2205	2869	(6.9)
Surinam	1567	1531	(3.7)
Venezuela	1290	1405	(3.4)
Totals	**8732**	**9434**	**(22.7)**

Centrally Planned	1989	1990	% 1990
China	1900	1700	(4.1)
Czecho-slovakia	205	209	(0.5)
E Germany	63	27	(0.1)
Hungary	891	848	(2.0)
Romania	611	400	(0.9)
USSR	4800	4000	(9.6)
Totals	**8470**	**7184**	**(17.2)**

Grand Totals 1989 - 41566
1990 - 41687

Figures refer to alumina hydrate.
Approximately 3.113 million tonnes of alumina production was used for non-metallic purposes in western countries in 1990 (3.148 million tonnes in 1989).

PRIMARY ALUMINIUM PRODUCTION, 1989-90
('000 tonnes and % of total 1990)

Developed	1989	1990	% 1990	Developing	1989	1990	% 1990	Centrally Planned	1989	1990	% 1990
Australia	1242.0	1232.7	(6.8)	Argentina	164.2	165.6	(0.9)	China	750.0	850.0	(4.8)
Austria	92.9	89.5	(0.5)	Bahrain	186.9	212.5	(1.2)	Czecho-			
Canada	1554.8	1567.4	(8.7)	Brazil	887.9	930.6	(5.1)	slovakia	69.3	69.8	(0.4)
France	334.9	325.9	(1.8)	Cameroon	87.3	87.5	(0.5)	East			
West				Egypt	179.5	179.6	(1.0)	Germany	53.9	21.0	(0.1)
Germany	742.0	720.3	(4.0)	Ghana	168.6	174.2	(1.0)	Hungary	75.2	75.2	(0.4)
Greece	148.3	149.7	(0.8)	India	423.3	433.2	(2.4)	N Korea	5.0	5.0	(-)
Iceland	88.7	87.8	(0.5)	Indonesia	196.9	192.1	(1.0)	Poland	47.8	46.0	(0.2)
Italy	219.5	231.8	(1.3)	Iran	45.0	59.4	(0.3)	Romania	269.1	177.8	(1.0)
Japan	35.0	34.2	(0.2)	S Korea	17.4	2.0	(-)	USSR	2500.0	2200.0	(12.2)
Netherlds	277.2	272.1	(1.5)	Mexico	71.7	67.5	(0.4)				
N Zealand	258.8	259.7	(1.5)	Surinam	28.4	31.3	(0.2)				
Norway	859.0	871.1	(4.8)	UAE	168.0	174.3	(1.0)				
S Africa	168.2	159.8	(0.9)	Venezuela	546.0	594.0	(3.3)				
Spain	352.5	355.3	(2.0)								
Sweden	97.0	96.3	(0.5)								
Switzer-											
land	71.4	71.8	(0.4)								
Turkey	61.8	60.9	(0.3)								
UK	297.3	289.8	(1.6)								
USA	4030.0	4048.3	(22.5)								
Yugo-											
slavia	342.1	350.5	2.0								
Totals	**11273.4**	**11274.9**	**(62.6)**		**3171.1**	**3303.8**	**(18.3)**		**3770.3**	**3444.8**	**(19.1)**
Grand Totals	1989 -	18214.8									
	1990 -	18023.5									

ALUMINIUM/BAUXITE/ALUMINA

MINE, REFINERY & SMELTER CAPACITIES, 1990
('000 tonnes) (Source: US Bureau of Mines)

	Bauxite	Alumina	Aluminium
Developed			
Australia	41391	11231	1233
Canada	-	1087	1567
France	490	606	326
W Germany	-	1165	720
Greece	2564	505	150
Ireland	-	927	-
Italy	-	752	232
Japan	-	890	34
Norway	-	-	871
New Zealand	-	-	260
Turkey	779	177	61
USA	495	5430	4048
Yugoslavia	2951	1086	351
Others	-	1133	1263
Total	**48610**	**25069**	**11116**
Developing			
Brazil	9876	1653	931
Ghana	381	-	174
Guinea	17524	642	-
Guyana	1424	-	-
India	4340	1334	432
Indonesia	1206	-	192
Jamaica	10937	2869	-
Sierra Leone	1445	-	-
Surinam	3267	1531	31
Venezuela	771	1425	594
Other Africa	7	-	428
Other Asia	501	-	448
Other Latin America	85	-	234
Total	**51764**	**9434**	**3465**
Centrally Planned			
China	4200	1700	850
Hungary	2559	848	75
USSR	5350	4000	2200
Others	230	636	320
Total	**12339**	**7184**	**3445**
TOTAL(a)	**112713**	**41687**	**18026**

(a) Other estimates show end 1990 bauxite capacity at almost 140 million tonnes, with non-metallurgical 7 million, alumina capacity at 44.8 million tonnes, with non-metallurgical over 3.4 million tonnes, and aluminium capacity at 20.75 million tonnes. (Source: JF King. World Capacity and Market Reports, November 1991.)

ALUMINIUM RECOVERED FROM SCRAP: WESTERN COUNTRIES
('000 tonnes 1989-90)

	1989	1990
European Community	1647	1588
Japan	999	1054
United States	2054	2393
Other Countries	683	699
Total	**5383**	**5734**

RESERVE PRODUCTION RATIOS FOR BAUXITE

Static Reserve Life (years)	220
Ratio of reserves to cumulative demand 1991-2010	14:1

CONSUMPTION OF PRIMARY ALUMINIUM

	000 tonnes		% p.a. growth rates		
	1989	1990	1960s	1970s	1980s
European Community	3913	4024	7.5	4.2	2.3
Japan	2204	2414	20.7	7.1	3.9
USA	4381	4325	7.8	2.8	-0.3
Others	4176	4138	4.7	7.5	4.4
Total Western World	**14674**	**14901**	**9.3**	**4.6**	**2.2**
Total World	**18122**	**17909**	**9.2**	**4.7**	**1.6**

END USE PATTERNS, 1990 %

Bauxite/Alumina (USA)

Aluminium metal : 92
Refractories, chemicals, abrasives and other products: 8

Aluminium	USA	Japan	W.Europe
Packaging	31	8	10
Building	19	25	20
Transport	22	30	16
Electrical	10	7	8
Consumer Durables	8	1	4
Mechanical	6	4	5
Others (inc Exports of Semis)	10	25	37

Sources:Metallgesellschaft and USBM

5

ALUMINIUM/BAUXITE/ALUMINA

VALUE OF CONTAINED METAL IN ANNUAL PRODUCTION

$24 billion (primary metal) at 1991 average LME price.

SUBSTITUTES

Bauxite/Alumina

Calcined clay can be substituted for refractory bauxite but only with reduction in length of life and in shock resistance. Sillimanite-alumina, silicon-carbide, magnesite-chromite and carbon-magnesite refractories are alternatives for high-alumina material but at higher cost. Silicon-carbide and diamonds can substitute for fused aluminium oxide in abrasive use but again at higher cost.

Aluminium

Plastics and steel compete in machinery, household appliances, and with glass and paper, in the container market, and magnesium, titanium and composites in the transport and structural industry. Wood is becoming increasingly important in the construction industry. Copper is used in many applications, and especially in electrical products.

Potential for substitutes often limited by relative weight (steel) or cost (titanium,magnesium).

TECHNICAL POSSIBILITIES

Bauxite/Alumina

Development of alternative raw materials including coal wastes, anthrosite, clay and shale is more dependent on political considerations than economics. No viable substitutes for aluminium production are in sight.

Possible development of other refractories, using nitrides and borides of titanium and zirconium.

Chemical use may be limited by development of chemicals or processes for recycling water.

Aluminium

Energy costs are a significant constraint on production in industrial countries. Advances in methods of alumina reduction should help keep aluminium competitive.

Development of composites and new alloys could reduce uses especially in transport applications.

Worries over health hazards could inhibit consumption in food packaging.

PRICES

Bauxite and Aluminium

Historically bauxite and alumina moved within integrated producers with pricing largely a book-keeping exercise. The diminishing importance of the integrated company and fundamental changes in the economic environment have resulted in more material moving under long term supply contracts. National bauxite levies and freight charges are major components of price. Spot purchases became more common in the alumina market from the mid 1980s mainly because of over supply. Prices of both bauxite and alumina are increasingly related to aluminium prices by percentage formulae.

Aluminium

	1986	1987	1988	1989	1990	1991
£/tonne						
LME cash standard grade	721.5-	758.75-	1076-			
Monthly average range	887.5	1309.5	23505			
$/tonne						
LME settlement high grade	-	-	1940-	1581-	1380-	1073-
Monthly average range	-	-	4280	2600	2245	1569
cents/lb						
LME standard grade	52.2	71.1	115.5	-	-	-
Real Dec 1991 prices	60.7	80.3	126.0	-	-	-
LME high grade	-		117.5	88.6	73.9	59.1
Real Dec 1991 prices	-	-	128.3	92.2	74.0	58.9

Prior to the late 1970s, pricing was dominated by long term producer contracts. The introduction of terminal markets (LME in 1979, Comex in 1983), plus structural changes in the market, mean that pricing is now far less rigid. The LME contract has flourished but the Comex contract has fallen into disuse. Today, a combination of flexible producer contracts and a dealer-controlled spot market exists, with LME pricing predominant. Producer contracts are short to medium term for fixed tonnages but with frequent price negotiations, linked to terminal market prices. Energy costs exert a strong influence on price, and conversely prices of energy for aluminium smelting, and sometimes even labour costs, are often linked to aluminium prices.

MARKETING ARRANGEMENTS

Bauxite and Alumina

A large proportion of sales is still within integrated producers, or through shared production arrangements, although there is an increasing trend towards independent smelters. The International Bauxite Association (IBA) has Jamaica, Guyana, Indonesia, Surinam, Ghana, Guinea, Sierra Leone, India, Yugoslavia and Australia as members. The IBA pools price and market information with the objective of 'fair and reasonable returns', although the search for minimum price arrangements has so far been unsuccessful.

ALUMINIUM/BAUXITE/ALUMINA

Aluminium

Substantial vertical integration from mine to fabricated product was a feature of the industry since the turn of the century with six companies and their associates dominating the stage: Alcan, Alcoa, Alusuisse, Kaiser, Reynolds and Pechiney. The oligopolistic nature of the industry has gradually been broken down since the early 1970s by the rise of independent smelters in energy rich nations of the Third World and Oceania. The difficult economic conditions of the early 1980s aided the process. Many of the large integrated concerns have now divested themselves of unprofitable subsidiaries although the 'big 6' still own some 40% of western world primary aluminium capacity. Through control over technology, their effective hold on the industry is even higher. The collapse of Comecon and the Soviet Union has brought new influences on world markets. Exports from the former Soviet Union rose considerably in 1991, mainly through traders, and helped depress prices. These exports seem likely to persist at high rates.

REAL PRICES 1979 to 1991
LME Aluminium

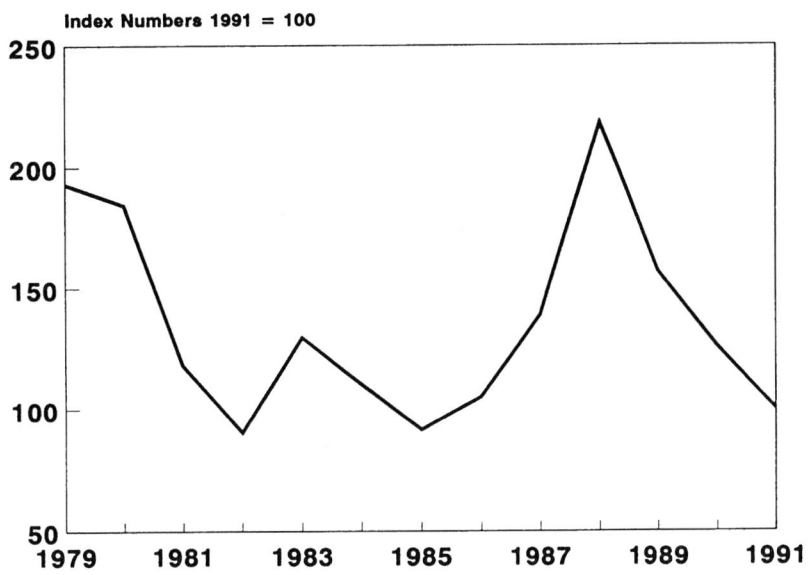

Index Numbers 1991 = 100

WORLD PRODUCTION 1979 to 1991
Primary Aluminium

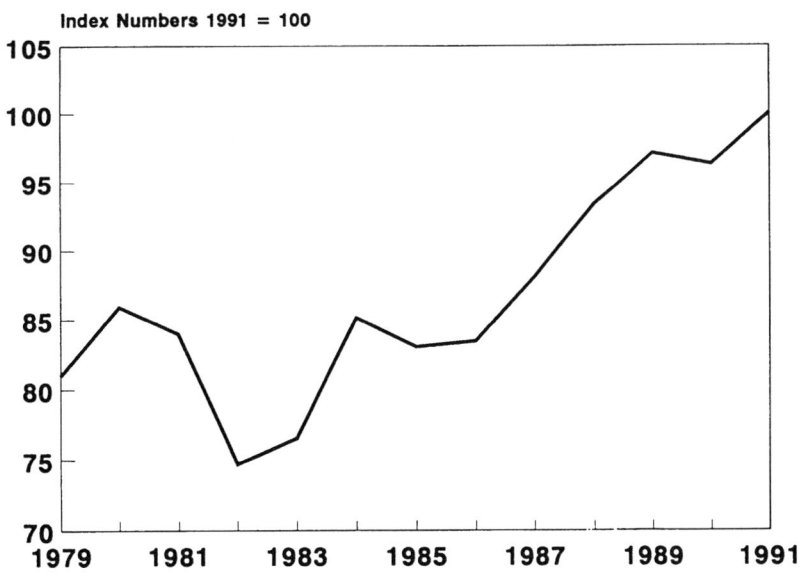

Index Numbers 1991 = 100

ALUMINIUM/BAUXITE/ALUMINA

SUPPLY AND DEMAND FOR BAUXITE BY MAIN MARKET AREA

	UK		EC(12)		Japan		USA	
	1989	1990	1989	1990	1989	1990	1989	1990
Production ('000 tonnes)	-	-	3334	2994	-	-	670	495
Net Imports ('000 tonnes)	313	317	9136	9898	2269	2302	12105	13817
Source of Net Imports (%)								
Australia	3	6	14	21	63	62	13	12
Brazil	-	7	1	2			11	14
China	4	8	4	4	4	3	3	2
European Community		11	6	-				
Ghana	80	71	4	3				
Guinea	-		66	60			29	28
Guyana	2	1	1	1	1	1	5	5
Indonesia	-		-		26	27	3	3
Jamaica	-		-				33	33
Malaysia	-		-		5	7	1	1
Sierra Leone	-	9	8				..	
Surinam	-		-					
Trinidad	-	-						
Others	-	1	1	1	1		2	2
Net Exports ('000 tonnes)	5.4	1.4	578	460	1.1	0.8	44	9
Consumption ('000 tonnes)	308	316 (apparent)	11892	12432 (apparent)	2268	2300 (apparent)	12730	14220 (apparent)
Import Dependence								
Imports as % of consumption	100	100	77	80	100	100	95	97
Imports as % of consumption & net exports	100	100	73	77	100	100	95	96
Share of World Consumption (%)								
Total World	..	..	11	11	2	2	12	13
Consumption Growth (% p.a.)								
1970s	6.0		1.1		6.6		-	
1980s	1.3		0.8		-6.5		-1.0	

SUPPLY AND DEMAND FOR ALUMINA BY MAIN MARKET AREA

	UK		EC(12)		Japan		USA	
	1989	1990	1989	1990	1989	1990	1989	1990
Production ('000 tonnes)	116	131	5009	5168	863	890	5180	5430
Net Imports ('000 tonnes)	655	625	1343	1350	28	19	4277	4035
Source of Net Imports (%)								
China					6	1		
Australia			19	20	12	28	84	85
Canada							3	3
European Community		32	48		16	16		
Brazil			1	1			1	2
USA				1	65	54		
Yugoslavia		2	2					
Guinea	5		11	1				
Jamaica	3		37	17			5	4
Surinam	2		26	30			5	4
Switzerland			3	3				
Others	58	52	1	25	1	1	2	2
Net Exports ('000 tonnes)	9	12	768	784	151	159	1295	1216
Consumption ('000 tonnes)	762	744 (apparent)	5584	5734 (apparent)	740	750 (apparent)	8162	8249 (apparent)
Import Dependence Imports as % of consumption	86	84	24	24	4	3	52	49
Imports as % of consumption and net exports	85	83	21	21	3	2	45	43
Share of World Consumption (%) Total World	2	2	13	14	2	2	20	20
Consumption Growth (%p.a.)								
1970s	6.0		-1.1		6.4		-	
1980s	0.1		0.2		-9.0		-1.6	

ALUMINIUM/BAUXITE/ALUMINA

SUPPLY AND DEMAND FOR ALUMINIUM METAL BY MAIN MARKET AREA

	UK		EC(12)		Japan		USA	
	1989	1990	1989	1990	1989	1990	1989	1990
Production ('000 tonnes)								
Primary Metal	297	290	2372	2345	35	34	4030	4048
Secondary Metal	220	221	1647	1588	999	1054	2054	2393
Total	**517**	**511**	**4019**	**3933**	**1034**	**1088**	**6084**	**6411**
Net Imports ('000 tonnes)	235	241	1597	1762	2363	2653	939	976
Source of Net Imports(%)								
European Community	16	8			2	1		
Iceland	9	9	3	3				
Norway	50	55	36	34				
Sweden		3	3	3				
Switzerland			2	1				
Yugoslavia		1	8	11				
Canada	7	7	4	4	6	6	85	85
USA	3	3	2	2	21	22		
S Africa			1		1	1		
Australia					24	20	1	1
New Zealand					8	8		
Cameroon			4	4				
Egypt			4	4				
Ghana	2	3	7	7			1	
Argentina			1	1	1	2	1	
Brazil	3	1	7	5	7	11	3	4
Surinam		3	1	1				
Venezuela	2	1	2	3	9	9	7	8
Bahrain					1	1		
UAE					4	4		
Indonesia					7	4		
Taiwan					1	1		
China						1		
Romania	2		5	2				
USSR	1	3	4	5	4	4		
Others	5	3	6	10	4	5	2	2
Net Exports ('000 tonnes)	169	157	233	267	3	5	655	798
Consumption ('000 tonnes)								
Primary Metal	455	454	3913	4024	2204	2414	4381	4325
Secondary Metal	205	181	1949	1935	854	887	2054	2393
Total	**660**	**635**	**5862**	**5959**	**3058**	**3301**	**6435**	**6718**

	UK		EC(12)		Japan		USA	
	1989	**1990**	**1989**	**1990**	**1989**	**1990**	**1989**	**1990**
Import Dependence								
Imports as % of consumption	36	38	27	30	77	80	15	15
Imports as % of consumption and net exports	28	30	26	28	77	80	13	13
Share of World Consumption(%)								
(Primary and Secondary Metal)								
Western World	3.3	3.0	29.0	21.5	15.1	15.8	31.8	32.1
Total World	2.7	2.5	23.8	23.6	12.4	13.1	26.1	26.6
Consumption Growth (% p.a.)								
1970s								
Primary Metal	1.6		4.2		7.1		2.8	
Total Metal	-1.6		4.1		7.3		3.2	
1980s								
Primary Metal	1.0		2.3		3.9		-0.3	
Total Metal	0.4		4.0		4.2		1.0	

ANTIMONY

WORLD RESERVES
('000 tonnes contained antimony and % of total)

Developed			Developing			Centrally Planned		
Australia	90	(2.1)	Bolivia	308	(7.3)	China	2180	(51.9)
Canada	65	(1.5)	Malaysia	120	(2.9)	Czecho-		
Italy	45	(1.1)	Mexico	181	(4.3)	slovakia	40	(0.9)
S Africa	236	(5.6)	Morocco	65	(1.5)	USSR	270	(6.4)
Turkey	90	(2.0)	Peru	66	(1.6)			
USA	82	(1.9)	Thailand	270	(6.4)			
Yugoslavia	90	(2.0)						
Totals	698	(16.2)		1010	(24.0)		2490	(59.2)
Grand Total		4200						

The world reserve base is 4.7 million tonnes and identified world resources are estimated at 5.0 million tonnes.

WORLD MINE PRODUCTION, 1989-90
('000 tonnes metal and % of total 1990)

Developed	1989	1990	% 1990	Developing	1989	1990	% 1990	Centrally Planned	1989	1990	% 1990
Australia	1.42	1.30	(2.4)	Bolivia	9.19	8.45	(15.8)	China	29.0	22.0	(41.0)
Austria	0.36	0.25	(0.5)	Guatemala	1.19	0.91	(1.7)	Czecho-			
Canada	2.82	0.74	(1.4)	Honduras	0.04	0.02	(-)	slovakia	0.79	1.10	(2.1)
								Romania	0.30	0.10	(2.1)
S Africa	5.20	5.26	(9.8)	Mexico	1.91	2.63	(4.9)	USSR	5.80	5.40	(10.1)
Turkey	1.46	1.50	(2.8)	Morocco	0.14	0.18	(0.3)	Others	0.11	0.10	(0.2)
USA(a)	2.50	2.30	(4.3)	Pakistan	0.03	0.04	(0.1)				
Yugoslavia	0.8	0.41	(0.7)	Peru	0.30	0.31	(0.6)				
				Thailand	0.49	0.54	(1.0)				
				Zimbabwe	0.14	0.06	(0.1)				
Totals	14.56	11.76	(21.9)		13.44	13.14	(24.5)		36.00	28.70	(53.6)
Grand Totals	1989 -	63.99									
	1990 -	53.60									

(a) From Metallgesellschaft. USBM states US mine output as negligible.

WORLD PRODUCTIVE CAPACITY, 1990

World production capacity is 113,000 tonnes.

SECONDARY PRODUCTION

Sizeable tonnages of antimony are contained in recycled antimonial lead, on which the available statistics are incomplete. In the United States secondary refineries produced 19,501 tonnes of contained antimony in 1989 and 20,380 tonnes in 1990. Approximately 17,000 tonnes was recovered from old scrap, largely as antimonial lead. In the United Kingdom total secondary recovery was 1483 tonnes per year in 1989 and 1990.

RESERVE/PRODUCTION RATIOS

Static Reserve Life (years):	78
Ratio of identified reserves to cumulative demand 1991-2010:	3.9 : 1

CONSUMPTION (Primary)

	tonnes		% p.a. growth rates	
	1989	**1990**	**1970s**	**1980s**
European Community	19000(b)	20475(b)	falling fast	n.a
United States	13424(a)	12739(a)	-3.2	1.9
Japan	14300(b)	14375(b)	-6.9	n.a

(a)Reported primary.
(b) Apparent metal & oxide (probably including secondary).

END USE PATTERNS, 1990 (USA) (%)

Flame retardants	70
Transport	10
Ceramics & Glass	4
Chemicals	10
Other	6

VALUE OF CONTAINED METAL IN ANNUAL PRODUCTION

$90 million (at average 1991 European Free Market metal price).

SUBSTITUTES

Tin, calcium, copper, selenium and cadmium are among the substitute hardeners for lead in batteries. Antimonial lead lost substantial market share in the early 1980s in batteries, mainly to low-maintenance and maintenance-free battery systems.

Antimony can be replaced by organic compounds or hydrated aluminium oxide in flame retardants and by tellurium and selenium in rubber manufacture.

Plastics or stainless steel products can replace enamel coated products. Titanium, zinc, chromium, tin and zirconium may be substituted in paints, pigments and enamels.

ANTIMONY

TECHNICAL POSSIBILITIES

Stabilisers in specialised plastics.

Advances in storage battery construction are bringing displacement of antimony and antimony recovery from this source is decreasing. Development of electric vehicles could utilise high-antimony batteries for deep-cycling characteristics.

Possible uses in aircraft night-vision systems and in space astronomy.

PRICES
(Source: Metal Bulletin)

	1986	1987	1988	1989	1990	1991
Ore Lump sulphide ore 60% Sb cif $/metric ton unit Sb. Range	19-23	19.5-25	20-26.5	16.5-21.5	15.3-19	15-17.5
Metal European Free Market Regulus 99.6% $/tonne	2580.0	2314.1	2181.9	1911.6	1702.2	1670.2
Real Dec 1991 prices	2994.0	2617.2	2384.3	1989.5	1721.9	1663.5

Changes in the supply/demand balance bring volatile prices. Fluctuating Chinese supplies ensure that the free market is most important although there has been some producer pricing for antimonial lead.

MARKETING ARRANGEMENTS

There is a mixture of state-owned production (Russia and China) and large private companies(e.g.: Consolidated Murchison in S Africa). Bolivia's mine output, once state-owned in Comibol, is privatised. Producers, consumers and traders belong to the Organizacion Internacional del Antimonio, which studies the problem of supply-demand imbalance, promotes the use of antimony and researches future uses and production techniques. China and Bolivia agreed in early 1989 to coordinate their policies on production and trade and issue producer prices, but Bolivia withdrew from the agreement in May 1990. Heavy Chinese exports of ores, concentrates and products have pushed world market prices down to the point where private sector mines elsewhere have been forced out of business. The USA imposed a 3.18% anti-dumping duty on Chinese antimony trioxide in October 1991.

REAL PRICES 1979 to 1991
Antimony European Free Market

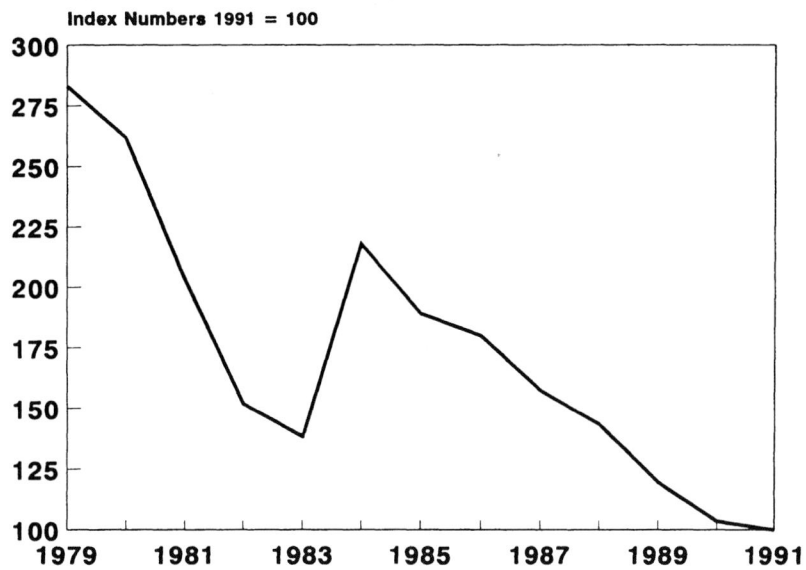

Index Numbers 1991 = 100

WORLD PRODUCTION 1979 to 1991
Antimony

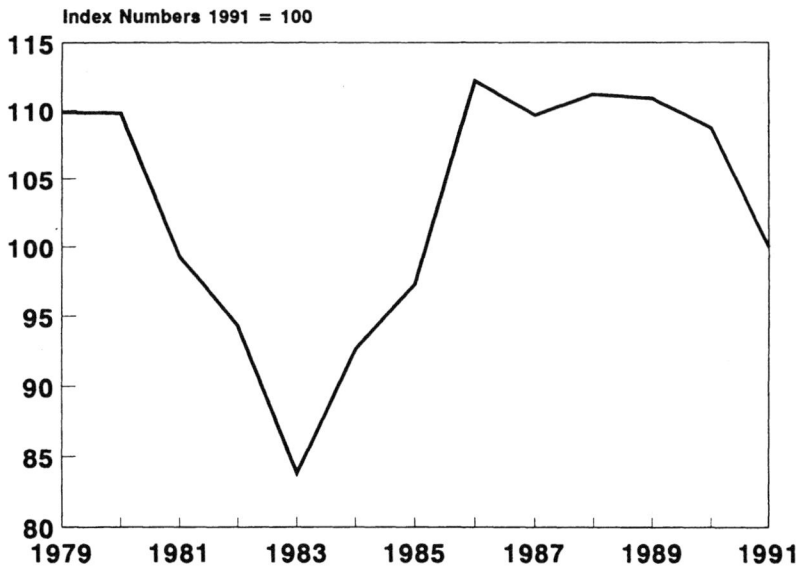

Index Numbers 1991 = 100

ANTIMONY

SUPPLY AND DEMAND BY MAIN MARKET AREA

	UK 1989	UK 1990	EC(12) 1989	EC(12) 1990	Japan 1989	Japan 1990	USA 1989	USA 1990
Production (tonnes)								
Mine (Sb content)	-	-	-	-	-	-	neg(a)	neg.(a)
Metal (primary)	n.a	n.a	n.a	n.a	173	216		
Oxides	n.a	n.a	n.a	n.a	10327	10994	18720	19085
							(Sb content)	(Sb content)
Residues						234	632	
Secondary production							19502	20380
								(SB content)

(a) USBM estimates

	UK 1989	UK 1990	EC(12) 1989	EC(12) 1990	Japan 1989	Japan 1990	USA 1989	USA 1990
Net Imports (tonnes) (gross)								
Ores and concentrates	n.a	n.a.	18872(a)	13535(a)	4978	5997	8613	5680
Sb content					4500	3454		
Unwrought metal	279	155	7381	9357	6204	6296	10057	13940
Oxide	1855	2151	4497	5879	7268	6299	12041	14472
Total Sb content	**n.a**	**n.a**	**211000**	**22495**	**14900**	**15200**	**25160**	**29480**
Wrought metal	178	450	586	1048				

(a) Excluding UK ores.

Source of Net Imports (%)

	UK 1989	UK 1990	EC(12) 1989	EC(12) 1990	Japan 1989	Japan 1990	USA 1989	USA 1990
Ores and Concentrates								
Australia			9		8	16		
Canada								1
S Africa			2					
USA								
Bolivia			36	32	48	35	15	18
Chile								1
China			39	46	42	48	19	49
Czechoslovakia			3	3				
Guatemala			4		2		13	8
Hong Kong							5	3
Mexico							44	16
Morocco								
Peru								3
Thailand								3
Turkey			3					
Zimbabwe								
Others			4	19		1	1	1

18

ANTIMONY

	UK 1989	UK 1990	EC(12) 1989	EC(12) 1990	Japan 1989	Japan 1990	USA 1989	USA 1990
Unwrought Metal (excluding alloys)								
European Community	22	20						2
Switzerland								1
China	76	33	71	76	98	97	92	85
Mexico								1
Hong Kong						1	3	1
Thailand			9	7	1	1	2	8
Turkey		39	5	3				
USSR			8	10				
Others	2	8	7	4	1	1	2	2
Oxide								
European Community	41	26			36	24	13	9
USA	3		5	3	1	3		
Mexico							8	29
Bolivia							6	5
China	45	72	80	85	57	63	42	35
Hong Kong						1	4	4
S Africa							26	18
Others	11	2	11	12	2	3	1	
Taiwan				4	4	6		
Net Exports (tonnes)								
Ores and concentrates	n.a	n.a	67(a)	179(a)	-	-	366	417
Unwrought metal	69	29	38(b)	44(b)	16	22	293	588
Oxides	c.5300	c.4700	7751(b)	6934(b)	713	958	2229	8605
Wrought metal	42	47	12	74	-	-	-	-

(a)excluding UK
(b)excluding Belgium/Luxembourg

Consumption (tonnes)								
	483(a)	481(a)	19000(d	20475(d)	670(a)	638(a)	13424(b)	12739(b) (reported)
	1483(e)	1483(e)			14300(c)	14375(c)	41499(f)	38963(f)

(a)Primary metal only including wrought (b)All primary forms Sb content (c)Metal + oxide (d)Mine output and Sb content of imports of ore, metals and oxide less exports (e)Scrap only (Sb content) (f) Apparent including secondary.

ANTIMONY

	UK		EC(12)		Japan		USA	
	1989	1990	1989	1990	1989	1990	1989	1990
Import Dependence (primary)								
Imports as % of consumption	100	100	100	100	100˙	100	100	100
Imports as % of consumption and net exports	100	100	100	100	100	100	100	100
Share of World Consumption (%)								
Total World Primary	n.a	n.a	n.a	n.a	n.a	n.a	21	24
Consumption Growth (% p.a.)								
1970s	n.a but large fall in 1970s		-7.5 (primary)		-6.9 (metal)		-3.2	
1980s	-2.1 (metal only)		n.a		n.a		1.9 (reported primary)	

ARSENIC

WORLD RESERVES

Arsenic is mostly found in association with deposits of complex base-metal ores,particularly copper-lead-zinc ores and arsenical pyrite copper ore. Arsenic trioxide is recovered as a byproduct during the smelting of such ores.

World reserves of arsenic, contained in copper and lead reserves, are estimated at 1 million tonnes. Half of these deposits are located in Chile (260,000t), USA (50,000t), Canada (50,000t),Mexico (60,000t), Peru (40,000t) and Philippines (40,000t), with the remainder principally in Europe (France and Sweden), Africa (Namibia) and Oceania.

The reserve base is 1.5 million tonnes and world resources contain approximately 11 million tonnes of arsenic. Arsenic trioxide = 76% contained arsenic.

ARSENIC

WORLD REFINERY PRODUCTION,1989-90, and PRODUCTION CAPACITY, 1990
(Arsenic trioxide tonnes and % of total 1990)

	Refinery Production		% of Production	Productive Capacity
	1989	1990	1989	1990
Developed				
Belgium	3500	3500	(7.3)	5000
Canada	2000	2000	(4.2)	4000
France	10000	7000	(14.7)	10000
W Germany	360	350	(0.7)	n.a
Japan	500	480	(1.0)	n.a
Portugal	180	160	(0.3)	n.a
Sweden	10000	10000	(21.0)	11000
Other Europe	n.a	n.a	n.a	n.a
Total	**26540**	**23490**	**(49.3)**	**30000**
Developing				
Bolivia	350	350	(0.7)	n.a
Chile	3400(a)	3400	(7.1)	8000
Mexico	5100(a)	4900	(10.3)	8000
Namibia	2900(a)	2000	(4.2)	3000
Peru	1000	600	(1.3)	2000
Philippines	5000	5090	(10.7)	8000
Total	**17750**	**16340**	**(34.3)**	**29000**
Centrally Planned				
USSR	8100	7800	(16.4)	10000
Total	**52390**	**47630**		**69000**

(a) Chile 4596, Mexico 7063, Namibia 2399, according to British Geological Survey. China is a major supplier of arsenic metal.
Recovery is also known to have occurred in recent years in Austria, China, Finland, Spain, UK, Yugoslavia and several East European nations.

RESERVES/PRODUCTION RATIOS

Static Reserve Life (years):   21
Ratio of identified reserve base to
cumulative demand 1991-2010:   1.1 : 1

CONSUMPTION

	'000 tonnes AS$_2$O$_3$		% p.a. growth rates	
	1989	1990	1970s	1980s
European Community	c.8	c.6.5	n.a	n.a
Japan	0.8	0.95	n.a	n.a
United States	29.6	27.3	-2.4	5.3

US consumption has been very volatile since peaking at 33,100 tonnes of arsenic trioxide in 1974.

END USE PATTERNS, 1990 (USA) (%)

Industrial Chemicals (wood preservatives and mineral flotation reagents)	70
Agricultural Chemicals (herbicides and plant desiccants)	22
Glass and Ceramics	4
Non ferrous alloys (metallic form)	2
Others (animal feed additives, pharmaceuticals, etc)	2

VALUE OF ANNUAL PRODUCTION

$25 million (arsenic trioxide at 1991 prices)

SUBSTITUTES

Substitutes exist in most end uses, although sometimes at higher cost. The imposition of increasingly tight environmental regulations encourages substitution.

A wide variety of organic compounds substitute for arsenical insecticides and herbicides. Creosote and pentachlorophenol are often interchangeable with the arsenical wood preservatives.

TECHNICAL POSSIBILITIES

Improvement in recovery of arsenic and controlling arsenic emissions during non-ferrous metal smelting. Increased environmentalist pressures against arsenic usage and disposal.

Replacement of silicon chip by gallium arsenide chip.

ARSENIC

PRICES

	1986	1987	1988	1989	1990	1991
Trioxide Mexican 99% AS_2O_3						
f.o.b. Laredo Texas (a) cents/lb	42.8	36.1	33.2	27	23	24
Real Dec 1991 prices	49.8	40.9	36.2	28.1	23	23.9
Metal Chinese(b) cents/lb	185	95	73	47	180	68

(a)Metals Week list price suspended after 1987, and 1987-91 prices calculated from US import statistics (average for imports from Mexico). (b) Based on US import data.

MARKETING ARRANGEMENTS

Refinery production of arsenic trioxide is dominated by a handful of large companies of which IMM in Mexico, Boliden in Sweden, Centromin in Peru and Penarroya in France are the most important. Namibia's output comes from Tsumeb Corporation.

REAL PRICES 1979 to 1991
Arsenic Trioxide. Mexican imported to USA

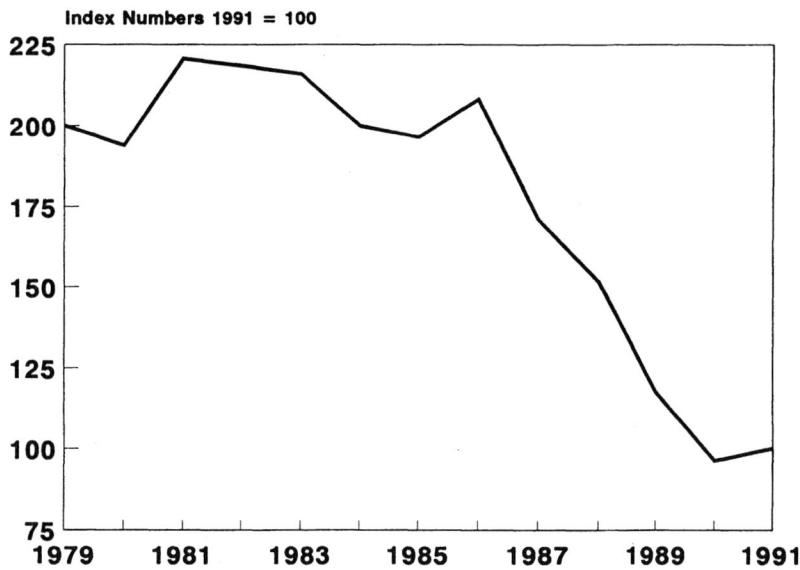

Index Numbers 1991 = 100

WORLD PRODUCTION 1979 to 1991
Arsenic

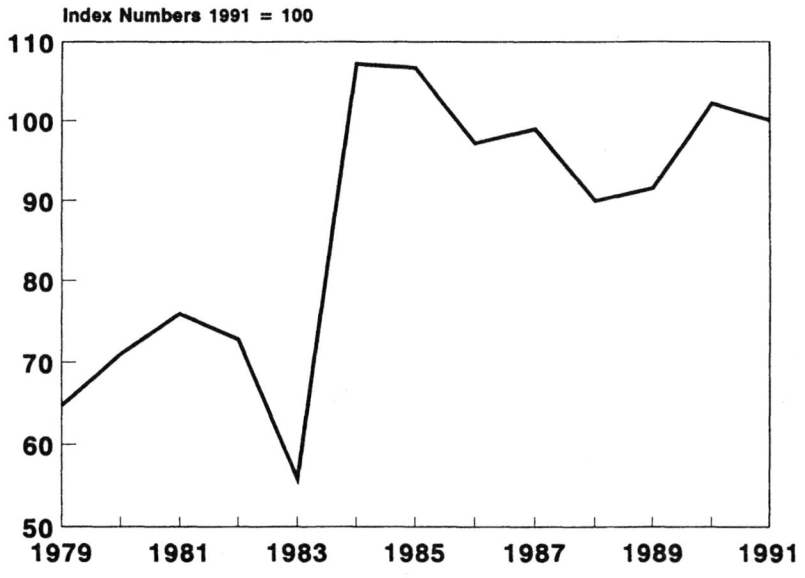

Index Numbers 1991 = 100

25

ARSENIC

SUPPLY AND DEMAND BY MAIN MARKET AREA

	UK 1989	UK 1990	EC(12) 1989	EC(12) 1990	Japan 1989	Japan 1990	USA 1989	USA 1990
Production (tonnes of arsenic trioxide)								
Mine Production	-	-	-	-	n.a	n.a	-	
Refinery	-	-	14040	11010	500	480	-	
Net Imports (tonnes)								
Arsenic trioxide	3779(a)	3238(a)	2892(a)	2898(a)	315	362	28347	26257
Metallic arsenic	70	80	567	631	30	95	928	796

(a)Includes sulphur trioxide

Source of Net Imports(%)

Arsenic Trioxide (c)

	UK 1989	UK 1990	EC(12) 1989	EC(12) 1990	Japan 1989	Japan 1990	USA 1989	USA 1990
Finland								4
Canada							6	
European Community	39	43			55	59	30	28
S. Africa	37		60					1
Sweden							20	9
Chile							18	23
China								4
Mexico							14	21
Philippines							10	10
S Korea					40	41		
Others	24	57	40	n.a	5		2	
Net Exports (tonnes)								
Arsenic trioxide	2939(a)	1903(a)	274(a,b)	140(a,b)	-	-	-	
Metallic arsenic	30	92	923	598	14	13	126	149

(a)Includes sulphur trioxide.
(b)Excludes Belgium-Luxembourg and France. Totals exceed 8677 in 1989 and 7565 in 1990.

Consumption

	UK 1989	UK 1990	EC(12) 1989	EC(12) 1990	Japan 1989	Japan 1990	USA 1989	USA 1990
(tonnes of arsenic trioxide)	n.a	n.a	c.8000	c.6500	c.835	c.950	29600	27345

	UK		EC(12)		Japan		USA	
	1989	1990	1989	1990	1989	1990	1989	1990
Import Dependence (Based on refined products)								
Imports as % of Consumption	100	100	47	58	43	51	100	100
Imports as % of Consumption and net exports	100	100	21	25	43	51	100	100
Share of World Consumption (%)								
Total World	n.a	n.a	c.15	c.14	c.2	c.2	c.56	c.57
Consumption Growth (% p.a.)								
1970s	n.a		n.a		n.a		-2.4	
1980s	n.a		n.a		n.a		5.3	

ASBESTOS

WORLD RESERVES
(million tonnes and % of total)

Developed			Developing			Centrally Planned		
Canada	40	(36.4)	Total	10	(9.1)	Total	50	(42.9
S Africa	5	(4.5)	(inc.Brazil			(inc.China		
USA	4	(3.6)	& Zimbabwe)			& USSR)		
Others	1	(0.9)						
(Australia, Cyprus,								
Japan, Yugoslavia)								
Totals	**50**	**(45.5)**		**10**	**(9.1)**		**50**	**(45.5**
Grand Total		**110**						

The reserve base is 143 million tonnes. The world's identified resources total 200 million tonnes, and hypothetical resources include an additional 45 million tonnes.

WORLD MINE PRODUCTION, 1989/90 and PRODUCTIVE CAPACITY, 1990
('000 tonnes and % of total 1990)

	Mine Production		% of Production	Productive Capacity
	1989	1990	1990	1990
Developed				
Canada	701	682	17.1	780
Cyprus	-	-	-	35
Greece	72	72	1.8	100
Italy	54	20	0.5	150
Japan	3	3	0.1	4
S Africa	155	147	3.7	470
Turkey	-	-	-	6
USA	17	20	-	30
Yugoslavia	7	7	0.2	20
Total	**1009**	**951**	**23.1**	**1595**
Developing				
Brazil	206	210	5.3	250
Colombia	8	8	0.2	20
India	36	37	0.9	30
Indonesia	-	-	-	n.a
S Korea	2	2	0.1	5
Swaziland	27	35	0.9	30
Zimbabwe	187	190	4.8	250
Others	1	1	-	1
Total	**467**	**483**	**12.2**	**586**
Centrally Planned				
China	160	160	4.0	170
USSR	2600	2400	60.4	2800
Total	**2760**	**2560**	**64.4**	**2970**
TOTAL	**4236**	**3992**		**5151**

Almost 94% of all asbestos mined is of crysotile. Amosite and crocidolite make up most of the remainder and are mined almost exclusively in South Africa. South African production capacity is 32% crysotile, 45% crocidolite and 23% amosite.

ASBESTOS

RESERVE/PRODUCTION RATIOS

Static Reserve Life (years):   27
Ratio of identified reserve base to
cumulative demand 1991-2010:   2.1 : 1

CONSUMPTION

	'000 tonnes		% p.a. growth rates	
	1989	1990	1970s	1980s
European Community				
(apparent)	310	275	1.1	-11.6
Japan (apparent)	298	291	1.9	0.4
United States	55	41	-3.9	-20.6

END USE PATTERNS, 1990 (USA) (%)

Asbestos-cement pipe and sheet	20
Friction products	26
Coating and compounds	5
Packing and gaskets	8
Paper	1
Roofing Products	37
Others	3

VALUE OF ANNUAL PRODUCTION

$1.5 billion approx. (based on 1991 average Canadian f.o.b. mine value)

SUBSTITUTES

Substitution is possible in many end uses, particularly asbestos-cement products where ceramic and new plastic materials are available, although few substitutes can give both physical and chemical characteristics at the same cost. Regardless of the technical difficulties involved, greatly tightened health regulations are hastening the replacement of asbestos in all uses in developed countries. Among the alternatives are glass-reinforced cement and artifical and natural fibres.

TECHNICAL POSSIBILITIES

Changes in manufacturing methods to reduce health hazards. The United States' Environmental Protection Agency banned the manufacture, import and processing of asbestos in July 1989, with full effect from August 1996.

Potential new uses of asbestos will be discouraged by environmental hazards. There is however some technical potential for use in high strength asphalt paving materials and as a reinforcing agent for lightweight plastics.

PRICES

	1986	1987	1988	1989	1990	1991
Canadian Chrysotile fibre $ C/short ton (range)						
Group 3 (Spinning fibre)	1550-2500	1550-2500	1550-2500	1550-2500	1550-2500	1550-2500
Group 4 (Shingle fibre)	1080-1500	1080-1500	1080-1500	1080-1500	1080-1500	1080-1500
Group 7 (Refuse/Shorts)	160-310	160-310	160-310	160-310	160-310	160-310
Group 7 US $/tonne	169.1	177.3	191.1	198.5	197.5	205.1
Real Dec 91 prices	196.4	200.4	208.7	206.5	198.1	204.4

Producer pricing in fixed contracts with discounting. Price depends on grade.

MARKETING ARRANGEMENTS

Asbestos is available in a number of different minerals but the majority of demand is for chrysotile. Vertical integration was a dominant feature in the industry, but this is fast breaking up with the tightening of environmental regulations, and health restrictions on production and use.

REAL PRICES 1979 to 1991
Asbestos Canadian Chrysotile Group 7

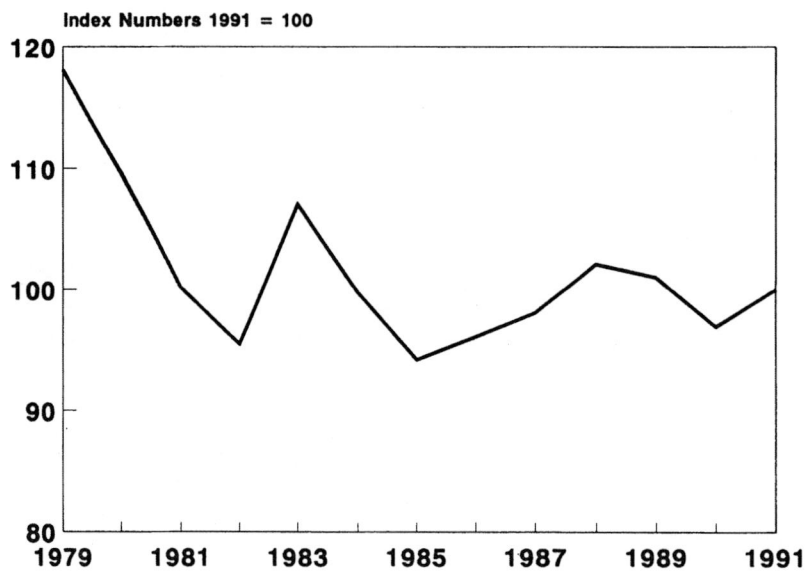

Index Numbers 1991 = 100

WORLD PRODUCTION 1979 to 1991
Asbestos

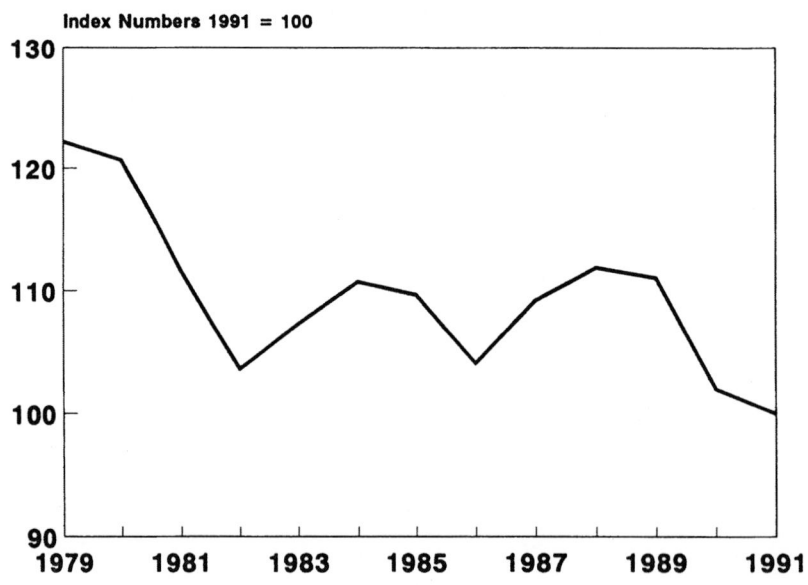

Index Numbers 1991 = 100

SUPPLY AND DEMAND BY MAIN MARKET AREA

	UK		EC(12)		Japan		USA	
	1989	1990	1989	1990	1989	1990	1989	1990
Production ('000 tonnes)	-	-	126	92	3	3	17	20
Net Imports ('000 tonnes)	19.7	16.0	250	239	295	288	55	41
of which crocidolite and amosite							0.6	0.8
Source of Net Imports(%)								
Canada	89	91	67	71	31	31	97	98
European Community	1				7	7		
S Africa	9		4	5	29	28	2	2
USA			1		4	5		
USSR			16	10	17	15		
Zimbabwe		7	10	9	12	12		
Others	1	2	2	5		2	1	
Net Exports ('000 tonnes)	0.3	0.3	66	56	0.1	0.1	27.0	28.0
Consumption ('000 tonnes)	19.4	15.7 (apparent)	310	275 (apparent)	298	291 (apparent)	55	41 (apparent)
Import Dependence								
Imports as % of consumption	100	100	81	87	100	100	100	100
Imports as % of consumption and net exports	100	100	67	72	100	100	67	59
Share of World Consumption (%)								
Western World	1	1	18	17	18	17	3	3
Total World	0.5	0.5	7	7	7	7	1	1
Consumption Growth (% p.a.)								
1970s	2.8		1.1		1.9			-3.9
1980s	-16.6		-11.6		0.4			-20.6

BARYTES

WORLD RESERVES
('000 tonnes and % of total)

Developed			Developing			Centrally Planned		
Canada	3000	(2)	Algeria	2000	(1)	China	40000	(24)
France	2000	(1)	Brazil	1000	(1)	USSR	10000	(6)
Germany	1000	(1)	Chile	1400	(1)	Others	n.a	
Greece	720	(..)	India	30000	(18)			
Ireland	1000	(1)	Mexico	7000	(4)			
Italy	2000	(1)	Morocco	10000	(6)			
Turkey	4000	(2)	Peru	2000	(1)			
USA	30000	(18)	Thailand	9000	(4)			
Yugoslavia	2000	(1)	Others	11000	(7)			
Others	720	(..)						
Totals	**46440**	**(27)**		**73400**	**(43)**		**50000**	**(30)**
Grand Totals		**170000**						

The world reserve base is 430 million tonnes and total world resources are believed to be roughly 2 bn. tonnes but only 500 million tonnes are identified.

WORLD MINE PRODUCTION, 1989-90
('000 tonnes and % of total 1990)

Developed	1989	1990	% 1990	Developing	1989	1990	% 1990	Centrally Planned	1989	1990	% 1990
Belgium	40	40	(0.7)	Algeria	60	60	(1.0)	China	1750	1750	(31.4)
Canada	39	48	(0.9)	Argentina	57	50	(0.9)	Czecho-			
France	100	100	(2.0)	Brazil	51	65	(1.1)	slovakia	60	60	(1.1)
W Germany	144	144	(2.6)	Chile	60	3	(0.1)	E Germany	30	27	(0.5)
Ireland	82	85	(1.5)	India	548	475	(8.5)	Poland	65	65	(1.2)
Italy	76	75	(1.3)	Iran	45	45	(0.8)	Romania	70	65	(1.2)
Japan	-	-	(-)	Mexico	325	332	(5.9)	USSR	540	500	(9)
Spain	6	7	(0.1)	Morocco	370	370	(6.6)				
Turkey	434	430	(7.7)	Thailand	76	80	(1.4)				
UK	70	67	(1.2)	Malaysia	36	47	(0.8)				
USA	290	439	(7.9)	Pakistan	28	30	(0.5)				
Yugoslavia	22	21	(0.4)	Tunisia	33	33	(0.6)				
Others	25	18	(0.3)	Others	44	41	(0.8)				
Totals	**1328**	**1473**	**(26.6)**		**1733**	**1632**	**(29.0)**		**2515**	**2467**	**(44.4)**
Grand Totals	**1989-**	**5576**									
	1990-	**5572**									

In addition, Bulgaria and Cuba are believed to produce Barytes.

PRODUCTIVE CAPACITY,1990
('000 tonnes)

Developed		Developing		Centrally Planned	
Canada	73	Algeria	109	China	1633
France	154	Morocco	544	Poland	100
W Germany	204	Brazil	154	Romania	82
Ireland	181	Chile	109	USSR	544
Italy	136	India	590	N. Korea	91
UK	91	Iran	181	Others	100
Turkey	408	Mexico	544		
USA	1451	Peru	91		
Others	147	Thailand	181		
		Others	276		
Totals	**2845**		**2779**		**2550**

Total World Capacity	**8174**

RESERVE PRODUCTION RATIOS

Static Reserve Life (years):	30
Ratio of identified reserve base to cumulative demand 1991-2010:	4.4 : 1

CONSUMPTION

	'000 tonnes		% p.a. growth rates	
	1989	1990	1970s	1980s
European Community (apparent)	792	837	-0.6	-
Japan (apparent)	129	116	0.4	2.4
United States	1277(a)	1434(a)	8.9	-7.1

(a)Ground and crushed barytes sold or used by processors. Apparent primary consumption is estimated at 1,271,000 tonnes in 1989 and 1,424,000 tonnes in 1990.

END USE PATTERNS, 1990 (USA) (%)

Drilling	85
Chemicals, Glass, Paint, Rubber	15

VALUE OF ANNUAL PRODUCTION

$0.5 billion (at 1991 average prices).

SUBSTITUTES

Drilling mud substitutes include celestite, iron ores, synthetic hematite and ilmenite, but the low costs and technical advantages of barytes deter substitution.

BARYTES

TECHNICAL POSSIBILITIES

Reclaiming and recycling of drilling muds would decrease the requirement for new supplies.

Increasing use in heavy concrete for radiation shields.

PRICES

	1986	1987	1988	1989	1990	1991
Drilling mud grade, Ground OCMA grade. Bulk delivered Aberdeen £/tonne (range)	45-57	45-50	45-50	45-50	45-50	45-52
Drilling mud grade $/tonne	78.3	77.9	84.6	77.8	84.9	85.8
Real Dec 1991 prices	90.7	87.9	92.3	80.9	85.0	85.5
Ground white paint grade 96-98%, $BaSO_4$ £/tonne (range)	125-135	125-135	135-185	135-185	135-185	135-185

Usually sold under long term supply contracts. Transport costs are important.

MARKETING ARRANGEMENTS

The barytes market is largely dependent upon the state of the oil and gas industry. The increasing importance of Chinese production has reduced the previous domination of the market by five US-based companies who together control or are associated with most of the major producing mines in the western world.

REAL PRICES 1979 to 1991
Barytes Drilling mud grade UK

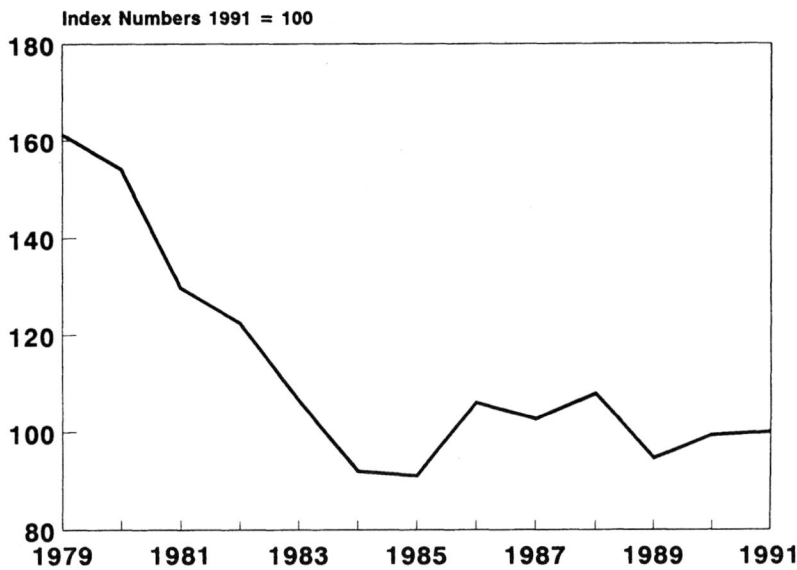

WORLD PRODUCTION 1979 to 1991
Barytes

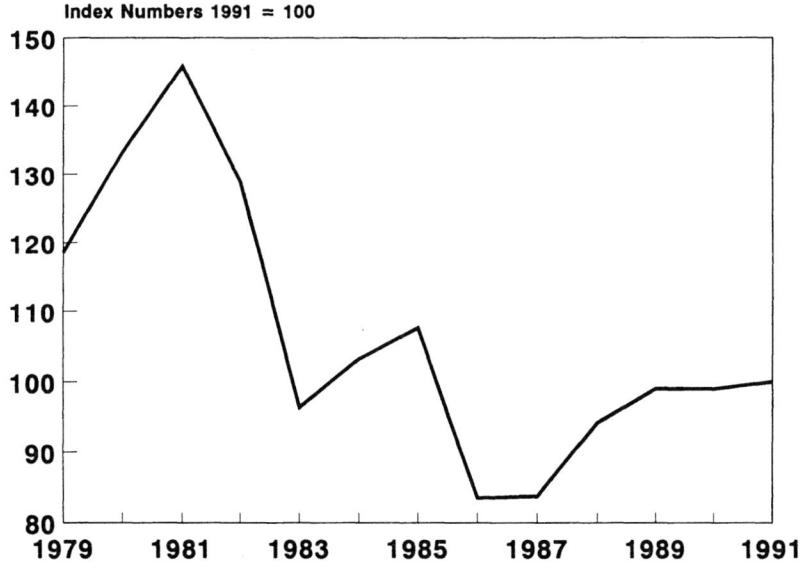

BARYTES

SUPPLY AND DEMAND BY MAIN MARKET AREA

	UK		EC(12)		Japan		USA	
	1989	1990	1989	1990	1989	1990	1989	1990
Production ('000 tonnes)	70	67.6	521	522	-	-	290	439
Net Imports ('000 tonnes)	113.0	217.4	368	413	1290	116	987.5 (crude)	987.5
							46.6 (ground)	56.7
Source of Net Imports (%)								
European Community	62	40						
Finland								
Norway		5		3				
Turkey			5	3				
Chile							2	
China	4	1	45	40	87	97	60	82
India			19	9	13	3	26	10
Mexico							5	4
Morocco	32	53	22	38				
Peru							4	3
Tunisia			8	7				
Others	2	1	1				3	1
Net Exports ('000 tonnes)	14.6	25.4	96.6	98.0	0.3	0.1	9.7	9.2
Consumption ('000 tonnes)	168	260 (apparent)	792	837 (apparent)	129	116 (apparent)	1277	1434
Import Dependence								
Imports as % of consumption	67	84	46	49	100	100	81	73
Imports as % of consumption and net exports	62	76	41	44	100	100	80	72
Share of World Consumption (%)								
Total World	3	5	14	15	2	2	23	26
Consumption Growth (% p.a.)								
1970s	6.8		-0.6		0.4		8.9	
1980s	4.7		-		2.4		-7.1	

BERYLLIUM

WORLD RESERVES
('000 tonnes of beryllium content and % of total)

Developed			Developing			Centrally Planned		
Australia	11	(2.9)	Argentina	25	(6.7)	China	n.a	(a)
Portugal	1	(0.3)	Brazil	140	(37.2)	USSR	61	(16.2)
S Africa	15	(4.0)	India	64	(17.0)			
USA	20	(5.3)	Mozambique	5	(1.3)			
			Rwanda	11	(2.9)			
			Uganda	15	(4.0)			
			Zaire	7	(1.9)			
			Zimbabwe	1	(0.3)			
Totals	**47**	**(12.5)**		**268**	**(71.3)**		**61**	**(16.2)**
Grand Total		**376**						

(a) Probably large.

Beryllium occurs in approximately 90 minerals with beryl and bertrandite as the two commercial ores. Only the USA has deposits of bertrandite. Outside the USA, firm data on beryl reserves are scarce because of the unpredictable nature of the concentration and occurrence of beryl. The data on non US reserves and resources are poorly delineated, and the figures above are, therefore, only broad estimates.

The reserve base is approximately twice the size of reserves, with additional deposits in Canada and Mexico.

BERYLLIUM

WORLD MINE PRODUCTION, 1989-90, and PRODUCTIVE CAPACITY, 1990
(tonnes of beryllium content and % of total 1990)

	Mine Production 1989	Mine Production 1990	% of Production 1990	Productive Capacity 1990
Developed				
Portugal	-	1	(0.3)	3
S Africa	-	-	-	3
USA	184	182	(53.9)	360
Total	**184**	**183**	**(54.2)**	**366**
Developing				
Argentina	2	3	(0.5)	4
Brazil	36	34	(9.5)	65
Madagascar	-	-	-	5
Mozambique	-	-	-	-
Rwanda	-	-	-	3
Zimbabwe	2	2	(0.5)	5
Total	**40**	**39**	**(10.5)**	**85**
Centrally Planned				
China	55	55	(14.8)	75
USSR	76	64	(20.5)	77
Total	**131**	**119**	**(35.3)**	**152**
TOTAL	**355**	**341**		**603**

Bolivia, Nepal and Namibia may also have produced beryl.

The beryllium produced came from about 9600 tonnes of beryl and bertrandite ores.

RESERVE/PRODUCTION RATIOS

Static Reserve Life (years):　　　　　}
Ratio of reserves to　　　　　　　　}　　very large
cumulative demand:　　　　　　　　}

CONSUMPTION

Consumption data are scarce except in the USA which is the world's major consumer. Its apparent consumption was 175 tonnes of contained beryllium in 1989 and 170 tonnes in 1990. US consumption declined by 1.7% per annum in the 1970s, and by 4.6% per annum in the 1980s.

END USE PATTERNS, 1990 (USA) (%)

Aerospace and defence applications (alloy & metal)	29
Electrical equipment (alloy & oxide)	19
Electronic components (alloy & oxide)	47
Other (alloy, oxide & metal)	5

VALUE OF CONTAINED METAL IN ANNUAL PRODUCTION

$220 million at average 1991 prices.

SUBSTITUTES

Although beryllium can be substituted in some applications, this usually results in substantial loss of performance. Its properties of light weight, high strength and high thermal conductivity preserve its markets, but they were hit from 1989 by concerns about potential health hazards.

Steel, titanium and graphite composites compete for structural uses of beryllium metal. Graphite is also an alternative for nuclear uses and in aircraft brake applications. Phosphor-bronze can sometimes be used in place of beryllium-copper alloys and sintered alumina in ceramic applications.

TECHNICAL POSSIBILITIES

Beryllium metal is likely to face competition from composites such as Graphite fibres.

Development of alternative ceramics such as aluminium nitride. New applications in low density aluminium alloys for aerospace and nuclear fields. Now uses have been developed in automotive electronics for a beryllium-copper alloy that contains less beryllium than the more common alloy.

PRICES

	1986	1987	1988	1989	1990	1991
Ore, imported into USA $/mtu	97.0	92.6	96.0	99.2	104.3	121.3
Ore, imported Real Dec 1991 price	112.6	104.6	104.7	103.3	104.4	120.8
US Domestic Metal Powder Blend $/lb	204	229	244	261	269	269

Ore contracts are usually negotiated on an individual basis. Metal prices are set by US producers.

MARKETING

Production is concentrated in USA, USSR, China and Brazil, with both the USA and USSR largely self sufficient. The USA is the world's major supplier of finished and semi-finished beryllium materials.

REAL PRICES 1979 to 1991
Beryllium Ore imported into USA

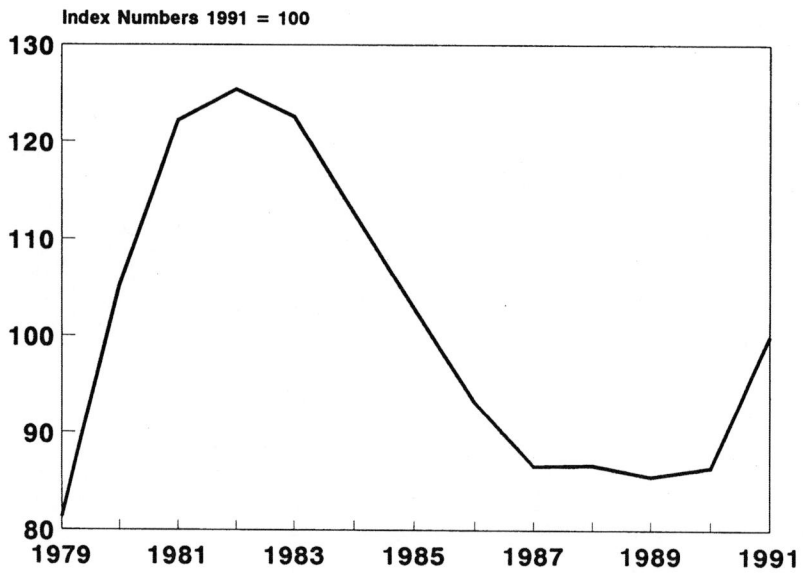

Index Numbers 1991 = 100

WORLD PRODUCTION 1979 to 1991
Beryllium ore

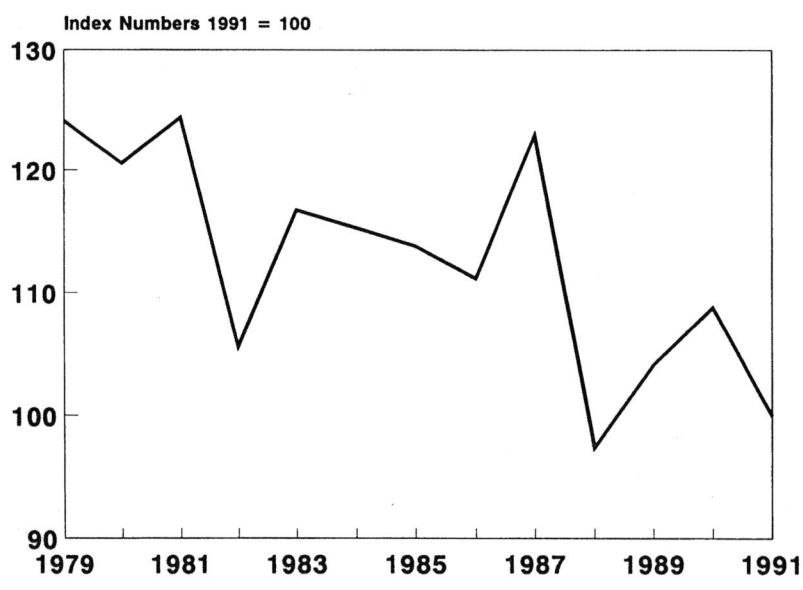

Index Numbers 1991 = 100

SUPPLY AND DEMAND BY MAIN MARKET AREA

	UK 1989	UK 1990	EC(12) 1989	EC(12) 1990	Japan 1989	Japan 1990	USA 1989	USA 1990
Production								
Mine	-	-	-	-	-	-	184	182
Imports (tonnes)								
Beryl ore(gross weight)							601	342
(contained beryllium)							24	14
Metal (unwrought)	5	3			2.0	0.1	14	11
Source of Net Imports (%)					Metal		Ore	Ore
USA					100	100		
Brazil							88	100
Others & unspecified							12	
Net Exports (tonnes)								
Metal (unwrought)	2	12	1	4	1	- (incl. alloys waste & scrap)	4	45
Consumption (tonnes) (a)Metal + oxide	n.a	n.a	n.a	n.a	n.a	n.a	187	175 (apparent)
Government stockpile acquisitions	-	-	-	-	-	-	6	21
Import Dependence								
Imports as % of consumption	100	100	100	100	100	100	20	14
Imports as % of consumption and net exports	100	100	100	100	100	100	20	11
Share of World Consumption (%)								
Total World	n.a	n.a	n.a	n.a	n.a	n.a	c.53(a)	c.51(a)

(a)Major western producer and consumer of primary beryllium products, metal, alloys and oxide.

Consumption Growth (%)								
1970s	n.a		n.a		n.a		-1.7	
1980s	n.a		n.a		n.a		4.6	

BISMUTH

WORLD RESERVES
('000 tonnes of contained bismuth and % of total)

Developed			Developing			Centrally Planned		
Australia	18	(16.8)	Bolivia	5	(4.7)	China	20	(18.7
Canada	5	(4.7)	S Korea	4	(3.7)	USSR	5	(4.7
Japan (metal)	9	(8.4)	Mexico	10	(9.3)			
US	9	(8.4)	Peru	11	(10.3)			
Other	11	(10.3)						
Totals	52	(48.6)		30	(28.0)		25	(23.4
Grand Total		107						

Bismuth is derived as a byproduct from various base metal ores, including lead, copper and tin; the above estimates of world reserves are based only on the bismuth content of lead and copper reserves. On the same basis, the reserve base is 250,000 tonnes. Coal ash is a potential source of bismuth, as are deep sea manganese nodules.

WORLD MINE PRODUCTION, 1989-90 and PRODUCTIVE CAPACITY, 1989
(tonnes of bismuth and % of total 1990)

	Mine Production		% of Production	Productive Capacity
	1989	1990	1990	1989
Developed				
Australia	400	400	(10.4)	1800(b)
Canada	157(a)	74(a)	(1.9)	700
Japan(c)	502	442	(11.5)	700
USA	100	100	(2.6)	700
Yugoslavia	40	9	(0.2)	150
Total	**1199**	**1025**	**(26.7)**	**4050**
Developing				
Bolivia	41	68	(1.8)	700(b)
S Korea	96	79	(2.1)	250
Mexico	883	717	(18.7)	1100
Peru	535	750	(19.5)	900
Total	**1555**	**1614**	**(42.1)**	**2950**
Centrally Planned				
China	786	1058	(27.6)	1000
Romania	60	60	(1.6)	100
USSR	80	80	(2.1)	100
Total	**926**	**1198**	**(31.2)**	**1200**
TOTAL	**3680**	**3837**		**8200**

(a) Includes content of exported concentrates.
(b) Includes mines on standby.
(c) Includes production from imported concentrates.

The production figures in the table are of recoverable bismuth in ores and concentrates.

In addition to the listed countries, Brazil, Bulgaria, Greece, E. Germany, Mozambique and Namibia are also believed to produce bismuth.

Belgium, W. Germany, Italy, and the United Kingdom have bismuth refinery capacity.

RESERVE/PRODUCTION RATIOS

Static Reserve Life (years):	28
Ratio of identified reserve base to cumulative demand 1991-2010:	3.6 : 1

BISMUTH

CONSUMPTION

	'000 tonnes		% p.a. growth rates	
	1989	1990	1970s	1980s
European Community	c.850	c.850	n.a	n.a
Japan (apparent)	894	587	n.a	n.a
USA (reported)	1352	1274	-0.6	1.8

END USE PATTERNS (%)

USA(1991)

Chemicals and pharmaceuticals	45
Fusible alloys	20
Metallurgical additives	33
Other alloys	1
Others (including experimental)	1

Japan(1989)

Catalysts	2
Pharmaceuticals	5
Fusible alloys	7
Metallurgical additives	18
Ferrites	52
Others	10

VALUE OF CONTAINED METAL IN ANNUAL PRODUCTION

$26 million (at average 1991 prices).

SUBSTITUTES

Antibiotics, magnesia and alumina are alternatives in pharmaceutical uses, mica and fish scales in cosmetics. Tellurium can substitute as a steel additive and plastics are an alternative for bismuth alloys in some casting applications. None of these substitutes is presently competitive. Indium can, however, compete in low temperature solders.

Bismuth's non-toxicity means that it continues to maintain most markets, and to expand into lead and cadmium markets on health grounds.

TECHNICAL POSSIBILITIES

New uses in bismuth-containing smoke and flame retardants, in electronic applications, plastic stabilisers, paint additives and batteries.

Increasing use of bismuth as an additive in free machining steel and to modify the carbon structure of ductile iron.

PRICES

	1986	1987	1988	1989	1990	1991
New York dealer 99.9 % min	3.14	3.67	5.78	5.74	3.54	3.02
New York dealer Real Dec 1991 prices	3.64	4.14	6.31	5.98	3.56	3.01
European Free Market min 99.99% $/lb	1.95-3.8	2.05-4.8	4.4-6.8	3.75-6.8	2.3-4.4	2.6-3.2

Source: Metal Bulletin.

Mostly producer pricing but dealer market has strong influence on prevailing price.

MARKETING ARRANGEMENTS

Present output is almost entirely from by-product sources (i.e. lead and copper). Most production shipped to major consuming countries for refining or direct use. Increasingly though, major producing countries are installing their own refining capacity. The role of China as both producer and consumer is rapidly increasing in importance.

REAL PRICES 1979 to 1991
Bismuth New York Dealer

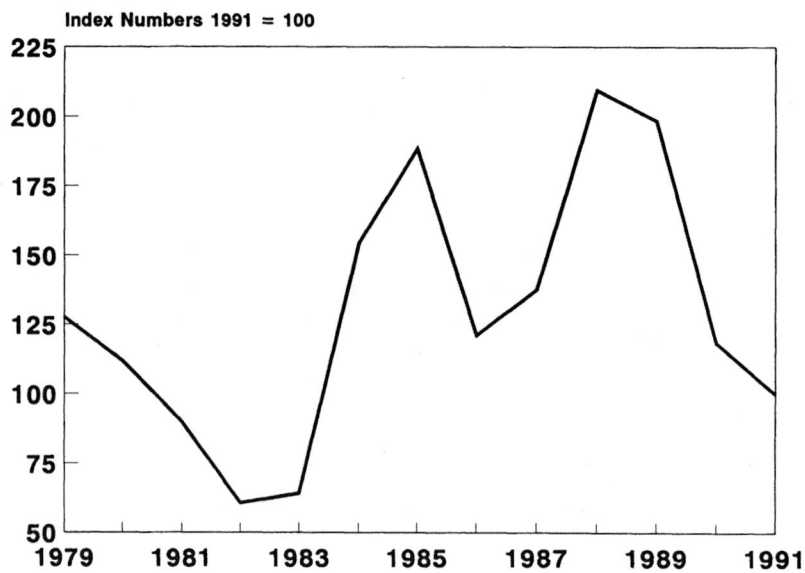

Index Numbers 1991 = 100

WORLD PRODUCTION 1979 to 1991
Bismuth

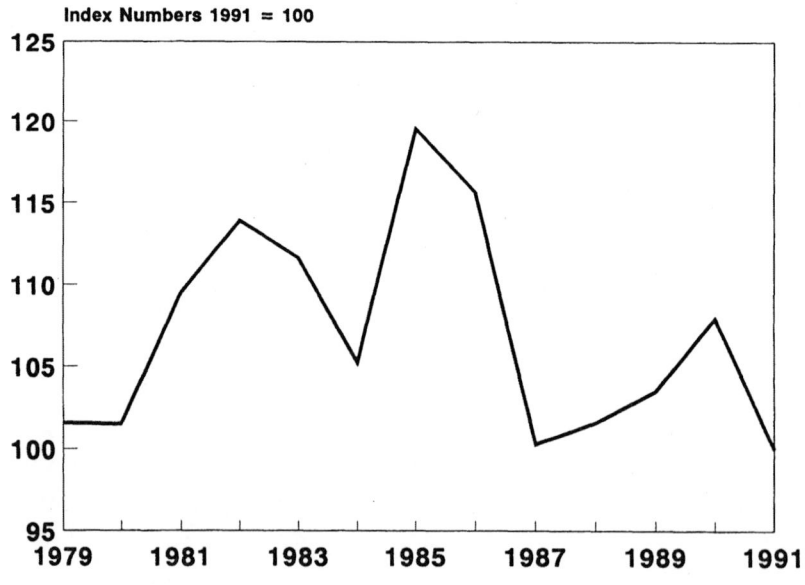

Index Numbers 1991 = 100

SUPPLY AND DEMAND BY MAIN MARKET AREA

	UK		EC(12)		Japan		USA	
	1989	**1990**	**1989**	**1990**	**1989**	**1990**	**1989**	**1990**
Production (tonnes)								
Mine	-	-	-	-	-	-	c.100	c.100
Metal	n.a	n.a	n.a	n.a	502	442	c.100	c.100
			(Italy produced 46 in 1989 and 34 in 1990)					
Imports (tonnes)								
Metal (incl alloys)	415	449	632(a)	591(a)	408	180	1880	1607

(a)excl. Belg/Lux

Source of Net Imports (%)

	UK		EC(12)		Japan		USA	
Australia	67		44					
Canada							8	
European Community	18	43			11	18	59	47
China			23		33		2	3
S Korea					23	65	1	
Bolivia							2	
Mexico		17		19	29	8	21	25
Peru	1	1	11	34	3	9	14	16
Others & undefined	14	39	22	47	1		1	1

Exports (tonnes)

	UK		EC(12)		Japan		USA	
Metal (incl alloys)	414	598	268(a)	214(a)	15.7	34.9	122	122

(a)excl. Belg/Lux.which exported 836 to USA in 1989

Consumption (tonnes)

	UK		EC(12)		Japan		USA	
Metal	n.a	n.a	c.850	c.850	894	587	1352	1274 (apparent)

Import Dependence (metal)

	UK		EC(12)		Japan		USA	
Imports as % of consumption	100	100	100	100	n.a	100	100	100
Imports as % of consumption and net exports	100	100	n.a	n.a	n.a.	38	100	98

Share of World Consumption (%)

	UK		EC(12)		Japan		USA	
Total World	n.a	n.a	c.23	c.22	c.24	c.15	c.37	c.33

Consumption Growth (% p.a.)

	UK		EC(12)		Japan		USA	
1970s	n.a		n.a		n.a		-0.6	
1980s	n.a		n.a		n.a		1.8	

BORON

WORLD RESERVES
(million tonnes of B_2O_3 content and % of total)

Developed			Developing			Centrally Planned		
Turkey	110	(31)	Argentina	3	(1)	China	27	(7.6
USA	136	(38)	Bolivia	5	(1.4)	USSR	54	(15.2
			Chile	13	(3.4)			
			Peru	6	(1.5)			
Totals	246	(69)		27	(7.5)		(81)	(22.8
Grand Total		354						

The B_2O_3 content of the total reserve base is 627 million tonnes.

Boron oxide (B_2O_3) = 31% contained boron.

WORLD MINE PRODUCTION, 1989-90, and PRODUCTIVE CAPACITY, 1990
('000 tonnes of B_2O_3 and % of total 1990)

	Mine Production		% of Production	Productive Capacity
	1989	1990	1990	1990
Developed				
Turkey	494	500	(41.4)	560
USA	562	608	(50.3)	735
Total	**1056**	**1108**	**(91.7)**	**1295**
Developing				
Argentina	33	33	(2.7)	28
Chile	25	25	(2.1)	30
Peru	2	2	(0.2)	5
Total	**60**	**60**	**(5.0)**	**63**
Centrally Planned				
China	5	5	(0.4)	5
USSR	40	35	(2.9)	41
Total	**45**	**40**	**(3.3)**	**46**
TOTAL	**1161**	**1208**		**1405**

RESERVE PRODUCTION RATIOS

Static Reserve Life (years):   295
(B_2O_3 content)
Ratio of reserve base to
cumulative demand 1991-2010:   23 : 1

CONSUMPTION

	'000 tonnes B_2O_3		% p.a. growth rates	
	1989	1990	1970s	1980s
European Community	c.400	c.400	n.a	n.a
Japan	c.80	c.80	1.2	n.a
USA	315	320	3.6	-0.8

END USE PATTERNS (%)

USA (1990)		Europe (1991)		Rest of World (1991)	
Glass products	58	Glass products	26	Glass products	37
Chemical fire		Ceramics	18	Ceramics	16
Retardants	4	Agriculture	2	Agriculture	8
Soap & detergents	8	Bleaches	39	Other	39
Agricultural &		Other	15		
biological	5				
Porcelain & enamel	3				
Other	22				

VALUE OF ANNUAL PRODUCTION

$0.9 billion (at average 1991 prices for contained B_2O_3).

SUBSTITUTES

Substitution is possible in most major uses, save glass products. Environmental concern may hasten substitution in soaps and detergents.

TECHNICAL POSSIBILITIES

Improvements in the evaporation of brine solutions are widening the choice of sources.
Production of boric acid through solution mining of colemanite. Substitution of borosilicate glass by plastic materials.
Re-formulation of detergents affects borate contents.

BORON

PRICES

	1986	1987	1988	1989	1990	1991
Borax, pentahydrate technical granular bulk ex works $ /tonne	243	249	249	272	272	272
Real Dec 1991 prices	281.1	281.2	271.0	282.8	272.3	270.9
Borax, pentahydrate UK. £/tonne	438.8	448	448	434.6	368.0	372.7

Products are sold principally under contract and list prices give an indication only.

MARKETING ARRANGEMENTS

Production is highly concentrated in the USA and Turkey. In the USA there are two producing companies, with one much larger than the other. The majority of Turkish output is controlled by the state owned Etibank. South American producers of brine-based boric acid, and more recently Russia, compete in marginal markets. The main markets are in the industrial countries.

REAL PRICES 1979 to 1991
Borax pentahydrate, bulk ex works

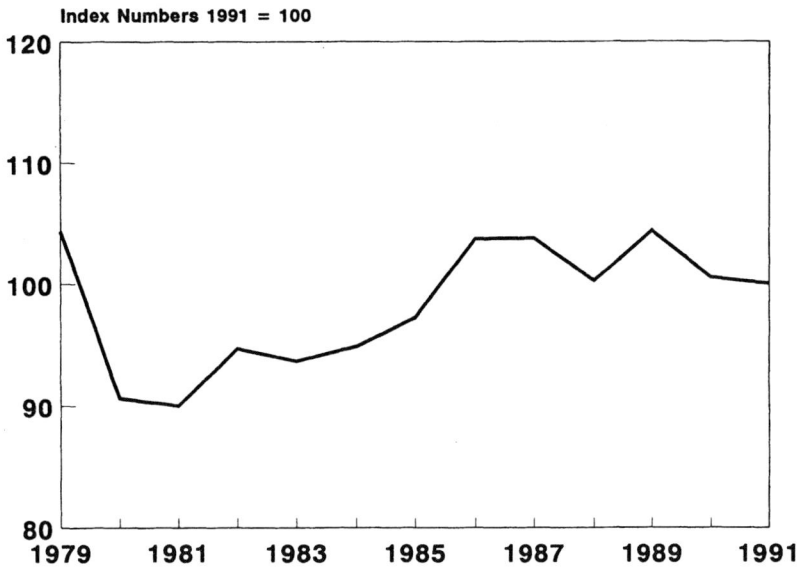

Index Numbers 1991 = 100

WORLD PRODUCTION 1979 to 1991
Borax

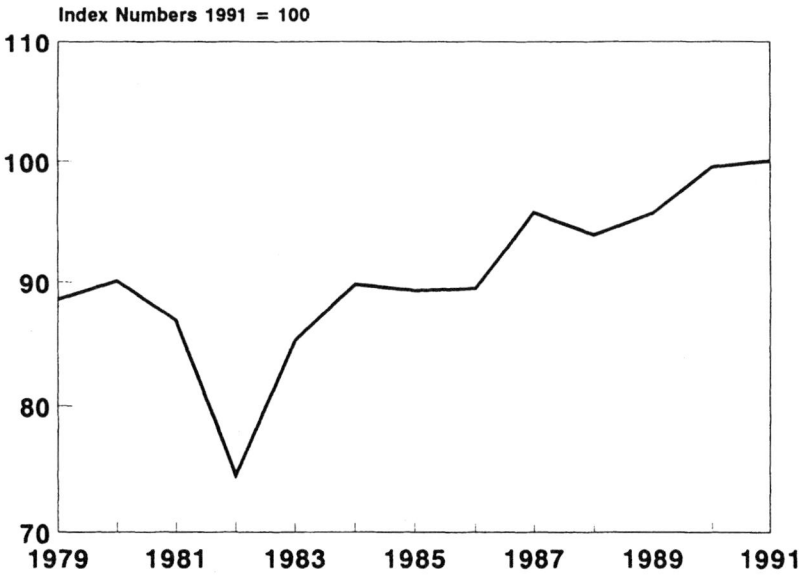

Index Numbers 1991 = 100

BORON

SUPPLY AND DEMAND BY MAIN MARKET AREA

	UK 1989	UK 1990	EC(12) 1989	EC(12) 1990	Japan 1989	Japan 1990	USA 1989	USA 1990
Production ('000 tonnes)								
Minerals	-	-	-	-	-	-	1114 562 (B_2O_3)	1094 608 (B_2O_3)
Net Imports ('000 tonnes)								
Minerals	47.7(a)	53.5(a)	489(a)	420(a)	65.9(a)	66.1(a)	15.0(a)	0.5(a)
Oxide and acid	13.5	11.8	21.7	31.7	28.1	26.6	2.9	5.6
(a)Includes crude natural boric acid.								
Source of Net Imports (%)								
Oxide and Acid								
EEC	92	79			18	16	58	53
USA	2	3	20	16	66	67		
China		1		3	1	1		
USSR				1	1	1		
Turkey	4	4	35	22	14	12		
Chile	1	7	17	30		3	37	45
Argentina	1	2	21	20				
Others & undefined	-	4	7	8			5	2
Ore								
EEC		2						
USA			7	3				
USSR					2	1		
Turkey	99	98	87	95	98	99	100	100
Others	1		6	2				
Net Exports ('000 tonnes)								
Oxide and acid	0.3	0.2	0.8(a)	1.4(a)	0.2	0.2	41.9	39.1
Refined sodium borates	8.0	7.6	15.6(b)	12.4(b)	0.2	0.2	555.1	585.0
Minerals	1.3	1.2	3.1	6.9	2.5	2.0	47.1	29.6

(a) excl. France and Italy
(b) excl. France, Belgium, Spain and Netherlands

	UK 1989	UK 1990	EC(12) 1989	EC(12) 1990	Japan 1989	Japan 1990	USA 1989	USA 1990
Consumption ('000 tonnes)	c.70	c.70	c.400	c.400	c.80	c.80	315 (B_2O_3)	320 (B_2O_3)

	UK		EC(12)		Japan		USA	
	1989	1990	1989	1990	1989	1990	1989	1990
Import Dependence								
Imports as % of consumption	100	100	100	100	100	100	-	-
Imports as % of consumption and net exports	100	100	100	100	100	100	-	-
Share of World Consumption (%)								
Total World	6	6	34	33	7	7	27	27
Consumption Growth (% p.a.)								
1970s	n.a		n.a		1.2		3.6	
1980s	n.a		n.a		n.a		-0.8	

CADMIUM

WORLD RESERVES
('000 tonnes of metal and % of total)

Developed			Developing			Centrally Planned		
Australia	55	(10.3)	Brazil	5	(0.9)	China	15	(2.8
Canada	80	(15.0)	India	15	(2.8)	Poland	10	(1.9
Ireland	15	(2.8)	Mexico	35	(6.5)	USSR	35	(6.5
Japan	10	(1.9)	Peru	25	(4.7)	Others	10	(1.9
S Africa	35	(6.5)	Zaire	20	(3.7)			
Spain	20	(3.7)	Others	50	(9.4)			
USA	70	(13.1)						
Others	30	(5.6)						
Totals	315	(58.9)		150	(28.0)		70	(13.1
Grand Total		535						

These figures are based primarily on estimated world resources of zinc. The world reserve base on the same basis is 970,000 tonnes and world resources exceed 6 million tonnes. Resources are substantially higher when allowance is made for other cadmium-bearing materials.

WORLD PRODUCTION OF REFINED CADMIUM AT SMELTERS, 1989-90
(tonnes of metal and % of total 1990)
Note:Cadmium is extracted from ores and concentrates, flue dusts and other materials, which sometimes include scrap. Statistics on mine production by country are not available.

Developed	1989	1990	% 1990	Developing	1989	1990	% 1990	Centrally Planned	1989	1990	% 1990
Australia	696	648	(3.2)	Algeria	46	65	(0.3)	Bulgaria	235	309	(1.5)
Austria	51	44	(0.2)	Argentina	60	55	(0.3)	China	874	1000	(5.0)
Belgium	1741	1956	(9.8)	Brazil	197	135	(0.7)	E Germany	26	17	(0.1)
Canada	1620	1437	(7.2)	India	275	277	(1.4)	N Korea	350	340	(1.7)
Finland	610	570	(2.9)	Mexico	1251	1207	(6.1)	Poland	555	373	(1.9)
France	170	187	(0.9)	Namibia	88	69	(0.3)	Romania	65	30	(0.2)
W Germany	1208	973	(4.9)	Peru	354	265	(1.3)	USSR	2600	2400	(12.0)
Italy	770	691	(0.9)	S Korea	470	568	(2.9)				
Japan	2694	2451	(12.3)	Zaire	224	127	(0.6)				
Netherlands	505	590	(3.0)								
Norway	207	286	(1.4)								
Spain	361	355	(1.8)								
Turkey	54	46	(0.2)								
UK	395	438	(2.2)								
USA	1550	1678	(8.4)								
Yugoslavia	476	362	(1.8)								
Totals	13108	12712	(63.7)		2965	2768	(13.9)		4705	4469	(22.4)
Grand Totals	1989- 1990-	20778 19949									

Includes secondary production where known.

REFINERY CAPACITY, 1990

World refinery capacity is 27,000 tonnes of which 15% is located in North America, 19% in Japan and a further 13% in Australia, Mexico and Peru. The remainder is mainly in Europe.

RESERVE PRODUCTION RATIOS

Static Reserve Life (years):   27
Ratio of identified reserve base to
cumulative demand 1991-2010
(based on zinc reserves alone):   2.1 : 1

CONSUMPTION

	tonnes		% p.a. growth rates	
	1989	1990	1970s	1980s
European Community	5870	6585	1.4	0.3
Japan(a)	3087	2612	-3.8	8.8
United States	4096	3107	-2.1	-1.3
Others	1820	1818	7.4	-
Total Western World(a)	**14873**	**14122**	**-**	**0.9**
Total World(a)	**18356**	**17181**	**0.7**	**0.3**

(a) Excluding stockpile purchases of 2028 in 1989 and 2216 in 1990.

END USE PATTERNS 1990(%)

USA

Coating, Plating	20
Pigments	16
Batteries	45
Plastics & synthetic products	12
Others (including alloys)	7

Japan

Batteries	74
Pigments	10
Alloys	4
Chemicals	1
Others	10

UK

Colours	59
Plating anodes & salts	14
Cadmium copper	-
Solder	3
Alloys	1
Miscellaneous (including batteries)	23

VALUE OF CONTAINED METAL IN ANNUAL PRODUCTION

$90 million (refined metal at average 1991 prices).

CADMIUM

SUBSTITUTES

The replacement of cadmium in all uses, but especially in pigments and plating, is being enforced by stringent existing and proposed regulations.

Zinc and aluminium can be substituted for some cadmium electroplating applications. Organotin compounds can be used in plastic stabilisers but at higher cost. Cadmium can be substituted in many alloys by a variety of metals, and inorganic compounds can replace it in paints and pigments. Lead-acid batteries can be used as a substitute for nickel-cadmium batteries but at the cost of reliability and longevity. Nickel hydride batteries hold greater promise.

TECHNICAL POSSIBILITIES

Solar energy cells, magnetic semiconductors and new forms of batteries.

Increased recovery from secondary sources and restriction on its use for environmental and health reasons could depress primary production. That will create problems in turn for zinc smelters, for which cadmium is an inevitable by-product.

PRICES

	1986	1987	1988	1989	1990	1991
European Free Market: (a) Ingots $/lb	0.92	1.77	7.03	6.13	3.12	2.02
Real Dec 1991 prices	1.07	2.0	7.67	6.39	3.12	2.01
US Producer Metal(b) 99.5% $/lb	1.26	1.99	8.94	-	-	-
New York dealer (Metals Week) $/lb	-	-	6.91	6.28	3.38	2.01

(a)Source: Metal Bulletin
(b)List price suspended in 1988

Producer pricing has largely given way to free market prices, although producers periodically exert control.

Cadmium is produced mainly as a by-product of zinc smelting and prices tend to bear little relationship to the supply/demand balance. At times of low prices, penalty clauses are sometimes imposed on zinc concentrates containing cadmium.

MARKETING ARRANGEMENTS

There is a wide spread of producers, with consumption largely concentrated in industrialised countries.

Environmental pressures are an increasingly important restraint especially in some industrialised countries.

Increased demand for use in nickel cadmium batteries is changing the patterns of the market.

REAL PRICES 1979 to 1991
Cadmium, European Free Market

Index Numbers 1991 = 100

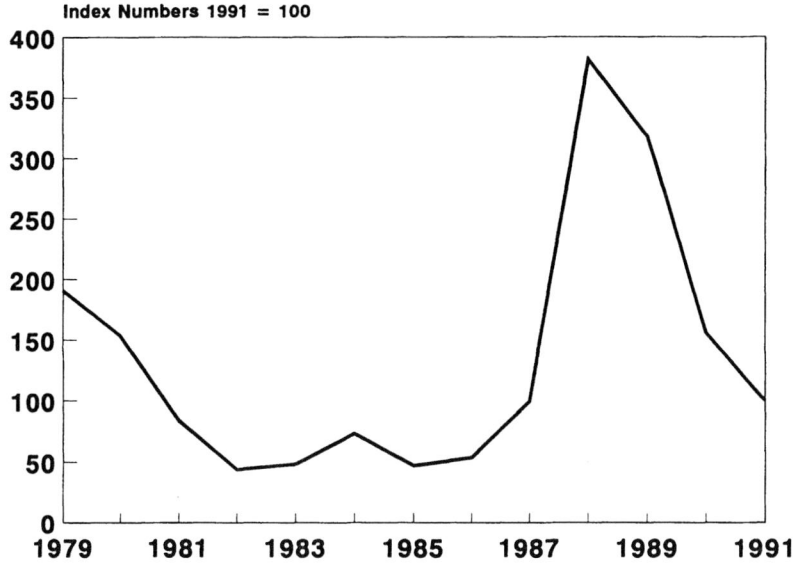

WORLD PRODUCTION 1979 to 1991
Cadmium, Refined Metal

Index Numbers 1991 = 100

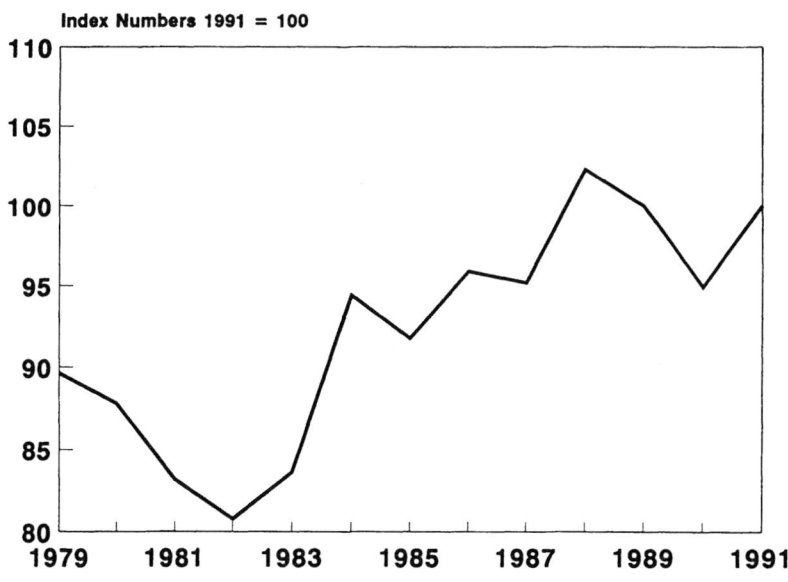

CADMIUM

SUPPLY AND DEMAND BY MAIN MARKET AREA

	UK 1989	UK 1990	EC(12) 1989	EC(12) 1990	Japan 1989	Japan 1990	USA 1989	USA 1990
Production (tonnes)								
Refined metal	395	438	5150	5190	2694	2451	1550	1678

Note: Production in the European Community and Japan is based mainly on imported ores.

	UK 1989	UK 1990	EC(12) 1989	EC(12) 1990	Japan 1989	Japan 1990	USA 1989	USA 1990
Net Imports (tonnes)								
Refined metal	762	812	2048	2046	2664	1946	2756	1714
Wrought metal	181	159	134	123	244	253	31	26
Source of Net Imports (%)								
Australia	3		4		14	10	6	11
Canada	5	7	11	7	12	16	34	43
European Community	44	56			34	20	27	17
Finland	28	19	24	12				
Norway	5	3	4	3			1	5
USA	7		7		4	1		
Yugoslavia								
Algeria								
China		1	10	9	9	4	1	1
Mexico			6				22	20
Peru					7	10	2	
S Korea					19	33	1	
Zaire	8	13	10	9				
Others and unidentified	-	1	24	60	1		6	3
Net Exports (tonnes)								
Refined metal	411	393	1582(a)	841(a)	118	202	369	385
Wrought metal	157	228	202	57	23	53	421	313

(a)Excluding W. Germany

	UK 1989	UK 1990	EC(12) 1989	EC(12) 1990	Japan 1989	Japan 1990	USA 1989	USA 1990
Consumption (tonnes)								
Refined metal	1332	934	5870	6585	3087(a)	2612(a)	4096	3107

(a) Excluding stockpile purchases of 2028 in 1989 and 2216 in 1990.

	UK		EC(12)		Japan		USA	
	1989	**1990**	**1989**	**1990**	**1989**	**1990**	**1989**	**1990**
Import Dependence*								
Imports as % of consumption	57	87	35	31	56(a)	51(a)	67	55
Imports as % of consumption	44	61	27	28	55(a)	43(a)	57	46

(a) Including stockpile purchases.

Share of World Consumption (%)								
Western world	9	7	40	47	21(a)	18(a)	27	22
Total world	7	5	32	38	17	15	22	11

(a) Excluding stockpile purchases.

Consumption Growth (% p.a.)				
1970s	-0.2	1.4	-3.8	-2.1
1980s	-3.3	0.3	8.8	-1.3

*Note: For the European Community and Japan this does not take account of the imported raw materials.

CHROMIUM

WORLD RESERVES
(million tonnes contained chromium and % of total)

Developed			Developing			Centrally Planned		
Finland	8.9	(2.1)	Brazil	2.3	(0.6)	Albania	1.9	(0.5)
S Africa	295.0	(70.5)	India	18.1	(4.3)	USSR	39.6	(9.5)
Turkey	2.5	(0.6)	Madagascar	2.1	(0.5)	Cuba	0.7	(0.2)
Greece	0.4	(0.1)	Philippines	2.3	(0.6)			
Japan	..	(..)	Zimbabwe	43.5	(10.4)			
			Others	1.3	(0.3)			
Totals	**306.8**	**(73.3)**		**69.5**	**(16.6)**		**42.2**	**(10.**
Grand Total		**418.9**						

The world reserve base totals approximately 2,100 million tonnes of contained chromium, 95% of which is found in S Africa and Zimbabwe. World resources total approximately 3,400 million tonnes contained chromium.

The above data assume a Cr_2O_3 content of 45% for chemical and metallurgical grade deposits and a 32% Cr_2O_3 content for refractory grade deposits. The former are usually classed as high-Cr and high Fe-chromite, the latter as high-alumina chromite.

WORLD MINE PRODUCTION, 1989-90
('000 tonnes gross weight and % of total 1990)

Developed	1989	1990	% 1990	Developing	1989	1990	% 1990	Centrally Planned	1989	1990	% 1990
Finland	498	500	(3.9)	Brazil	476	476	(3.7)	Albania	610	600	(4.6)
Greece	56	60	(0.5)	India	1003	995	(7.7)	Cuba	51	50	(0.5)
Japan	12	8	(..)	Iran	62	60	(0.5)	USSR	3800	3800	(29.5)
S Africa	4951	4498	(34.9)	Madagascar	63	62	(0.5)	Vietnam	3	3	(..)
Turkey	1000	850	(6.6)	New Caledonia	60	60	(0.5)	China	25	25	(..)
Yugoslavia	12	11	(..)	Pakistan	27	20	(..)				
				Philippines	173	198	(1.5)				
				Sudan	25	15	(...)				
				Zimbabwe	627	600	(4.6)				
				Oman	-	-	(..)				
				Indonesia	8	8	(..)				
Totals	**6529**	**5927**	**(45.9)**		**2524**	**2494**	**(19.5)**		**4489**	**4478**	**(34.6)**
Grand Totals	**1989-**	**13542**									
	1990-	**12899**									

In addition, Bulgaria, North Korea and Thailand are believed to produce chromite. Assuming an average 44% Cr_2O_3 content, the chromium content of mine production was 4 million tonnes in 1989 and 3.8 million tonnes in 1990.

World ferrochromium production was 3.5 million tonnes in 1989 (chromium content almost 2 million tonnes) and 3.36 million tonnes in 1990 (chromium content 1.9 million tonnes).

WORLD MINE CAPACITY 1990
('000 tonnes of contained chromium)

Developed		Developing		Centrally Planned	
Finland	211	Brazil	108	Albania	218
Greece	21	India	241	Cuba	14
Japan	3	Iran	15	USSR	1100
S Africa	1505	Madagascar	45	Vietnam	1
Turkey	440	New Caledonia	-	China	13
		Pakistan	2		
		Philippines	60		
		Sudan	2		
		Zimbabwe	169		
		Indonesia	20		
Totals	2180		662		1346
Grand Total	4188				

RESERVE/PRODUCTION RATIOS

Static Reserve Life (years): 105
Ratio of identified reserve base to
cumulative demand 1991-2010: 23 : 1 approx.(4.5:1 for reserves)

CONSUMPTION

	'000 tonnes		% p.a. growth rates	
	1989	1990	1970s	1980s
European Community	970	750	6.5	0.7
Japan	745	609	4.1	2.9
United States	378(a)	335(a)	0.5	-4.6

The figures cover the chrome content of all forms.

(a) 452 in 1989 and 447 in 1990 including secondary.

END USE PATTERNS 1990 (USA) (%)

Chromite:Intermediate Outlets
Metallurgical and chemical	
industry	89
Refractory industry	11

Ferroalloys and metal
Stainless and heat-resisting steels	78
Other steels	11
Superalloys	3
Cast irons	2
Other	6

CHROMIUM

VALUE OF ANNUAL PRODUCTION

$0.7 billion (as chromite at 1991 average price).

SUBSTITUTES

Substitutes deterred by cost, performance or customer appeal for chromium. There are no substitutes in stainless steel or superalloys.

Boron, manganese, nickel and molybdenum can be substituted in alloy steels and cast irons. Base metal alloys can sometimes be used in place of stainless steel.

Dolomite is an alternative for some refractory bricks. Cadmium yellow is one of several alternative pigments, and nickel and zinc ores are possible substitutes for decorative coating protection.

TECHNICAL POSSIBILITIES

Changing steel technology is reducing the use of chromite refractories.

PRICES

	1986	1987	1988	1989	1990	1991
Ore, Transvaal 44% Cr_2O_3 no ratio $/tonne	41.0	42.5	47.7	62.0	59.2	50.1
Ore, Transvaal Real Dec 1991 prices	47.6	48.0	52.0	64.5	59.5	50.0
Ore, Turkish 48% Cr_2O_3 3:1 ratio $/tonne	125	106.8	140.9	205.3	155.3	129.2
Metal, US Electrolytic 99.1% Cr $/lb	3.54	3.45	3.54	3.75	3.89	3.97
Ferrochrome, US low C (.05% C) imported dealer price cents/lb	83.7	86.8	123.3	108.3	106.6	98.7
Ferrochrome charge 50-55% US imported cents/lb	39.9	43.8	85.1	74.7	47.5	47.3

Most ore is sold on long term contracts but there is a small free market.

MARKETING ARRANGEMENTS

Ore production is increasingly highly concentrated, with large state (e.g. former USSR, Etibank in Turkey) and private interests (e.g. S Africa). Some ore producers are linked with ferroalloy companies but only two firms (Outokumpu Oy of Finland and Samancor of S Africa) are totally vertically integrated from chromite mining to stainless steel production. There is a growing trend towards steel industry use of lower grade ferrochrome and towards production of ferrochrome near mines. Ferrochrome production in USA and Europe has become increasingly uncompetitive. The South African industry is undergoing concentration with a recent merger of two large producers, raising fears amongst consumers about future security of supplies.

REAL PRICES 1979 to 1991
Chrome Ore, Transvaal 44% Cr2o3

Index Numbers 1991 = 100

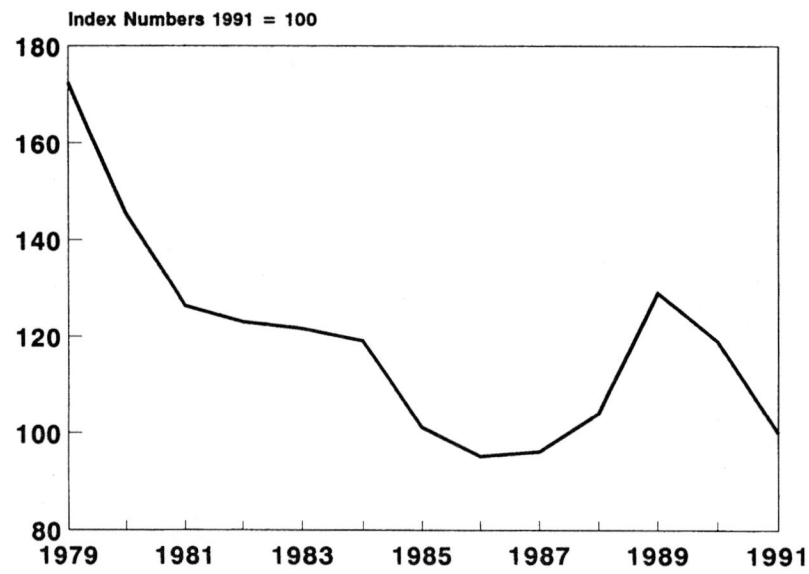

WORLD PRODUCTION 1979 to 1991
Chrome ore

Index Numbers 1991 = 100

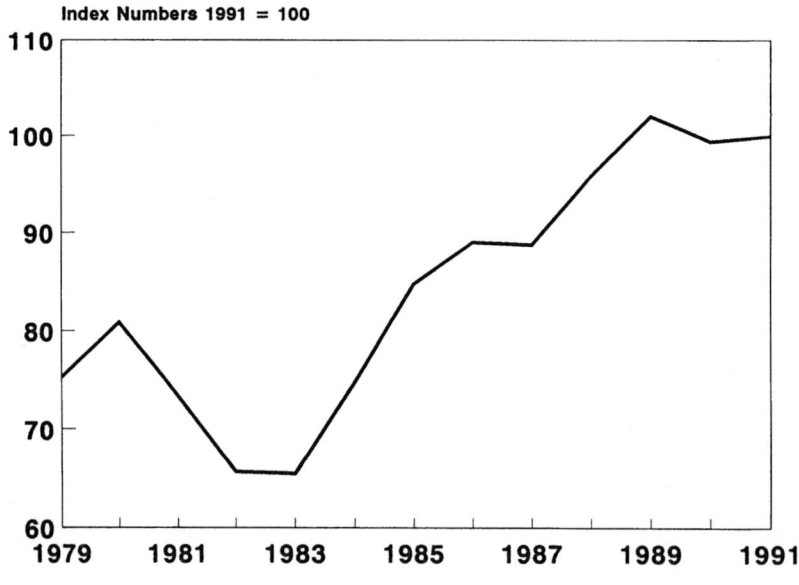

SUPPLY AND DEMAND BY MAIN MARKET AREA

	UK 1989	UK 1990	EC(12) 1989	EC(12) 1990	Japan 1989	Japan 1990	USA 1989	USA 1990
Production								
('000 tonnes)								
Chromite Ore (gross)	-	56		60	12	8	-	-
Ferrochromes (gross) and metal	c.100	c.100	310	293	324	293	147	109
			(excl. metal)		(excl. metal)			
Ferrochrome and metal (Cr content)	c.57	c.57	175	167	185	167	90.1	67.7
			(excl. metal)		(excl. metal)			
Net Imports								
('000 tonnes)								
Chromite ore (gross)	170	155	1126	795	1043	789	525	306
Ferrochromes (gross)	112	67	745	687	477	442	344	416
Chromium Metal(incl.alloys)	0.51	0.67	2.01	2.49	0.8	0.6	4.1	6.7
Source of Net Imports (%)								
Chromite Ore								
European Community	4	5						
Finland				1				
Norway			2					
S Africa	95	93	44	49	57	68	70	85
Turkey			17	13	6	4	20	4
Albania			23	23	7	5		
USSR			12	6	2	3	1	
India					11	3		
Madagascar					7	7		
New Caledonia					3	3		
Philippines			2	3	2	4	4	4
Central Africa				1				
Pakistan						2		
Iran				2	3			
Others and unidentified	1	2		2	2	3	2	2

CHROMIUM

	UK 1989	UK 1990	EC(12) 1989	EC(12) 1990	Japan 1989	Japan 1990	USA 1989	USA 1990
Ferrochrome (= 58% contained chromium)								
European Community					1	1	4	4
Finland			7	6				6
Norway				2				5
S Africa			58	63	59	62	40	41
Sweden			4	5			4	3
Turkey			1				12	11
Yugoslavia	n.a.	n.a.	4	2			9	12
China			1	1	3	2	6	
USSR			7	3	2	2		
Albania			3	3		1		
Brazil			2	1	3	1	2	
India			3		10	10	1	
Philippines			1		11	11	5	3
Zimbabwe			9	11	9	7	15	13
Others & unidentified			-	3	2	3	2	2

Net Exports
('000 tonnes)

	UK 1989	UK 1990	EC(12) 1989	EC(12) 1990	Japan 1989	Japan 1990	USA 1989	USA 1990
Chromite ore (gross)	0.1	-	26.4	18.6 (a)	2.4	0.5	40.4	6.3
Ferrochrome (gross)	0.8	0.8	35.4(a)	29.0	1.1	2.8	7.0	8.3
Chromium metal incl.alloys	3.5	3.3	4.2	3.8	2.3	2.8	1.5	0.3

(a)Excl. Germany.

Consumption
('000 tonnes)

	UK 1989	UK 1990	EC(12) 1989	EC(12) 1990	Japan 1989	Japan 1990	USA 1989	USA 1990
Chromite ore (gross)	170	155	1256	836	1053	797	561	402
Ferrochromes (gross)	211	166	1020	951	800	732	360	389
Total consumption (Cr content)	140	107 (apparent)	970	750 (apparent)	745	609 (apparent)	378	335 (reported)

(a)Apparent, including secondary 452 in 1989 and 447 in 1990.

Import Dependence (chromite)

	UK 1989	UK 1990	EC(12) 1989	EC(12) 1990	Japan 1989	Japan 1990	USA 1989	USA 1990
Imports as % of consumption	100	100	97	96	99	99	100	100
Imports as % of consumption and net exports	100	100	96	96	99	99	100	100

Share of World Consumption (%)

	UK 1989	UK 1990	EC(12) 1989	EC(12) 1990	Japan 1989	Japan 1990	USA 1989	USA 1990
Total World	3	3	24	20	19	16	10	9

Consumption Growth (% p.a.)

	UK	EC(12)	Japan	USA
1970s	6.5	-4.8	4.1	0.5
1980s	n/a	0.7	2.9	-4.6

COBALT

WORLD RESERVES
('000 tonnes of contained cobalt and % of total)

Developed			Developing			Centrally Planned		
Australia	23	(0.7)	Brazil	15	(0.5)	USSR	140	(4.2)
Canada	45	(1.4)	Botswana	5	(0.2)	Cuba	1040	(31.3)
Finland	23	(0.7)	Guatemala	10	(0.3)			
S Africa	20	(0.6)	India	18	(0.5)			
Yugoslavia	10	(0.3)	Indonesia	25	(0.7)			
			New Caledonia	230	(6.8)			
			Zaire	1360	(40.9)			
			Zambia	360	(10.8)			
			Zimbabwe	2	(0.1			
Totals	**121**	**(3.7)**		**2025**	**(60.8)**		**1180**	**(35.5)**
Grand Total		**3326**						

The world's estimated reserve base is 8.3 million tonnes. In addition to the above countries, there are deposits in the USA, Peru, Morocco, Philippines, Uganda, Papua New Guinea and Albania. Identified world resources total 11 million tonnes of cobalt with millions of tonnes of potential resources also contained in seabed nodules.

COBALT

WORLD MINE AND METAL PRODUCTION, 1989/90, and PRODUCTIVE CAPACITIES, 1990
(tonnes of metal and % of total)

	Mine Production 1989	1990	% of Production 1990	Metal(a) Production 1989	1990	% of Production 1990	Productive Capacity 1990
Developed							
Australia	1000	1000	(2.7)	-	-		-
Canada	2340	2290	(6.1)	2123	2184	(8.5)	3000
Finland	-	-	-	1295	1109	(4.3)	1800
France	-	-	-	160	170	(0.7)	6000
Japan	-	-	-	102	199	(0.8)	2800
Norway	-	-	-	1946	1831	(7.1)	2000
S Africa	730	730	(2.0)	525	525	(2.0)	1000
USA	-	-	-	-	-	-	900
Total	**4070**	**4020**	**(10.8)**	**6151**	**6018**	**(23.5)**	**12100**
Developing							
Morocco	121	120	(0.3)				
Botswana	215	205	(0.6)	-	-	-	-
Brazil	200	200	(0.8)	30	60	(0.2)	-
New Caledonia	800	800	(2.2)	-	-	-	-
Philippines	-	-	-	-	-	-	-
Zaire	25000	20000	(53.8)	9311	10033	(35.1)	18000
Zambia	7255	7100	(19.1)	4490	4644	(18.7)	5000
Zimbabwe	90	90	(0.2)	100	100	(0.4)	-
Total	**33681**	**28515**	**(76.7)**	**13931**	**14827**	**(57.9)**	**23000**
Centrally Planned							
Albania	600	600	(1.6)	10	20	(0.1)	-
China	-	-	-	270	270	(1.1)	-
Cuba	1825	1600	(4.3)	-	-	-	-
USSR	2850	2400	(6.5)	5300	4500	(17.5)	6600
Total	**5275**	**4600**	**(12.4)**	**5380**	**4790**	**(18.7)**	**6600**
TOTAL	**43026**	**37135**		**25662**	**25645**		**41700**

(a) Including cobalt content of cobalt salts.

Much Zairean production is further processed in Belgium.
According to the British Geological Survey mine output was only 32600 tonnes in 1989, with a different geographical split from the USBM's estimates shown above.

A number of other countries mine cobalt-containing ores but data are inadequate for reliable estimates to be made. For many countries the figures cover the content of ore raised rather than cobalt recovered. There is often a considerable discrepancy between them. There are large cobalt-rich waste dumps in Zaire, Zambia and Uganda from which cobalt could be recovered if market conditions were favourable.

RESERVE/PRODUCTION RATIOS
Static Reserve Life (years): 90 (land only)
Ratio of identified reserve base to
cumulative demand 1991-2010: 18 : 1 (land only)

CONSUMPTION

	tonnes		% p.a. growth rates	
	1989	1990	1970s	1980s
European Community	c.6000	c.6800	0.2	n/a
Japan	2585	2994	0.7	2.7
				(metal only)
United States	7152(a)	7472(a)	-	0.7

(a)Reported consumption. Apparent consumption is higher at 7164 in 1989 and 7885 in 1990, excluding GSA purchases.

END USE PATTERNS, 1990

USA

Superalloys	40
Magnetic alloys	10
Cutting & wear-resistant materials	10
Chemical & ceramic use	10
Others (mainly alloy steels, non-ferrous alloys & welding materials)	20
Catalysts	10

Japan

Speciality steels	30
Magnetic alloys	22
Cutting materials	.11
Catalysts	14
Tubes, sheets, rods, wires	13
Others	10

VALUE OF CONTAINED METAL IN ANNUAL PRODUCTION

$0.95 billion (refined metal at 1991 free market price).

SUBSTITUTES

There are few effective substitutes for cobalt in most major end uses.The continuing trend is towards reduction of, rather than elimination of cobalt in alloys,eg: iron-base, heat-resistant alloys for cobalt-base materials in turbine applications. Ceramic parts appear increasingly competitive in high-performance uses.

In less demanding applications, nickel- and ferrite-magnets are among the alternatives for permanent magnets. In catalytic applications, molybdenum and aluminium are complements and nickel and tungsten together are substitutes. Nickel, vanadium, chromium or tungsten alloys may, in time, replace those containing cobalt as the binder in cemented carbides. Nickel may be substituted for cobalt in several applications but often only with a loss of effectiveness.

TECHNICAL POSSIBILITIES

Exploitation of cobalt-bearing manganese nodules from the deep sea, during the next century. Recovery of cobalt from tailings, dumps. Improved scrap recovery.

Use of cobalt alloy coatings on video recording tape and on computer diskettes.

Substitution of ceramic components for those currently fabricated from superalloys.

COBALT

PRICES

	1986	1987	1988	1989	1990	1991
Metal European Free Market 99.5% Co $/lb	7.0	6.5	7.1	7.55	9.98	16.77
Real Dec 1991 prices	8.09	7.33	7.71	7.85	9.99	17.31

Source: Metal Bulletin

Cobalt is mainly produced as a by-product of copper or nickel and output is relatively independent of supply/demand balances. Until 1981 prices were mainly producer contracts, but slack demand led to substantial discounting, and from 1982 the market was dominated by spot purchases at merchants' terms. Producers reasserted their control in early 1984 but it broke down again in 1986 before being restored in 1987. The Central African producers have since maintained market discipline, but at the expense of building up large stocks at times. Political disturbances, especially in Africa, can have a dramatic effect on price as witnessed most recently in 1990-91.

MARKETING ARRANGEMENTS

Zaire's Gecamines is the major producer and could once strongly influence price and supply through a varying production, even though cobalt is a by-product, and by stockpiling. Technical and economic difficulties in Zaire's mines reduced Gecamines' flexibility in 1990-91, greatly contributing to the rise in prices. USSR and Cuba are important producers and Canadian nickel producers make sizeable sales. The Cobalt Development Institute promotes the use of cobalt and serves as an information centre for the metal.

Cobalt is regarded as a strategic metal in many uses, and this led the USA to upgrade its stockpile holdings in 1989-90.

REAL PRICES 1979 to 1991
Cobalt, European Free Market, Metal

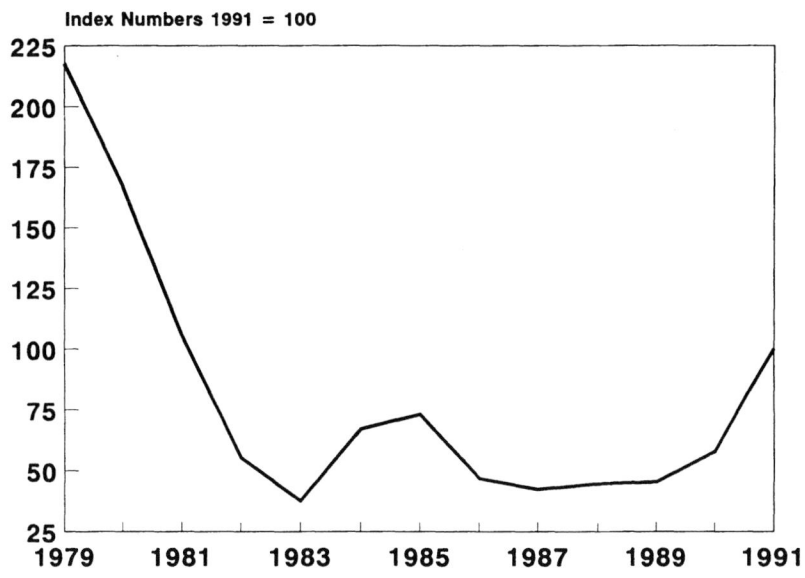

WORLD PRODUCTION 1979 to 1991
Cobalt Metal

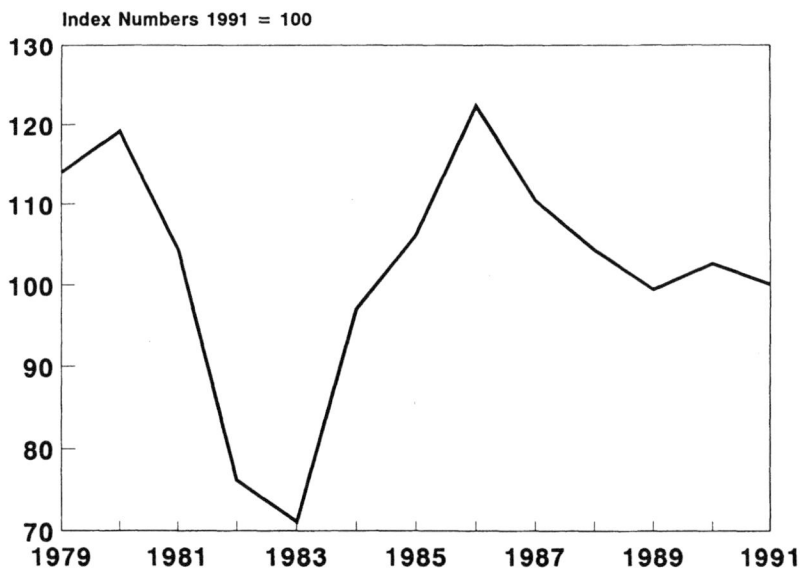

COBALT

SUPPLY AND DEMAND IN MAIN MARKET AREAS

	UK 1989	UK 1990	EC(12) 1989	EC(12) 1990	Japan 1989	Japan 1990	USA 1989	USA 1990
Production (tonnes)								
Mine output	-	-	-	-	-	-	-	-
Secondary recovery	n.a	n.a	n.a	n.a	n.a	n.a	1184	1225
Primary metal (inc. salts)	n.a	n.a	160	170	102	199	-	-
			(exc. UK Belgian processing of imported materials)		(from imported ores & matte from Australia, Philippines & N Caledonia)			
Net Imports (tonnes)								
Metal (unwrought)& scrap	2279	3222	5345(a)	6260(a)	3354	4642	5952	6322
Oxide	501	543	1084	1230	449	631	380	488
Other forms(wrought)	335	547	502(a)	748(a)	22	165	59	47
Total all above forms (Co content)	2975	4160	c.6630	c.7900	3700	5260	5793	6529

(a)Excludes Belgium-Luxembourg

Source of Net Imports (%)

Metal and Oxide

	UK 1989	UK 1990	EC(12) 1989	EC(12) 1990	Japan 1989	Japan 1990	USA 1989	USA 1990
Canada	23	15	4	10	2	1	21	10
European Community		13	20		33	27	19	52
Finland	4	3	10	11	3	3	2	2
Japan							1	1
Norway	7	3	11	7	6	7	11	6
S Africa			2				4	8
Switzerland	3		2					
USA	6	10	7	8	3	4		
Zaire	15	15	30	31	42	38	18	7
Zambia	18	11	16	13	10	19	22	12
USSR		2	1	3				
Tanzania	5	5	2	3				
Namibia		9		4				
Sweden		4		3				
Others	6	3	5	7	1	1	1	2

	UK		EC(12)		Japan		USA	
	1989	1990	1989	1990	1989	1990	1989	1990
Net Exports (tonnes)								
Metal (unwrought)& scrap	673	876	517(a)	493(a)	356	439	353	520
Oxides	1200	1414	544(a)	762(a)	21	99	603	922

(a) Excludes Belgium-Luxembourg from which USA imported 909 (Co content) in 1989 and 388 (Co content) in 1990.

	UK		EC(12)		Japan		USA	
Consumption (tonnes)								
All forms Co content	c.1200	c.1900	c.6000	c.6800			7152 (reported)(a)	7472
					(metal) 2585	2994	7164 (apparent)	7885(a)

(a) Includes secondary.

	UK		EC(12)		Japan		USA	
Import Dependence								
Imports as % of consumption (exc. scrap)	100	100	100	100	100	100	100	100
Imports as % of consumption and net exports (exc. scrap)	100	100	100	100	100	100	100	100
Share of World Consumption (%)								
Western World Primary (approx.)	5	7	23	26	10+	12+	23	26
Consumption Growth (% p.a.)								
1970s	-0.5		0.2		0.7		-	
1980s	n.a		n.a		2.7		0.7	

COPPER

WORLD RESERVES
(million tonnes of contained copper and % of total)

Developed			Developing			Centrally Planned		
Australia	7	(2.2)	Chile	85	(26.5)	Poland	10	(3.1
Canada	12	(3.7)	Mexico	14	(4.4)	USSR	37	(11.5
Portugal	3	(0.9)	Papua New			Mongolia	3	(0.9
S Africa	2	(0.6)	Guinea	7	(2.2)	Others	2	(0.6
USA	55	(17.1)	Peru	8	(2.5)			
Others	8	(2.5)	Philippines	10	(3.1)			
			Zaire	26	(8.1)			
			Zambia	12	(3.7)			
			Others	20	(6.2)			
Totals	87	(27.1)		182	(56.7)		52	(16.2
Grand Total		321						

The reserve base is 522 million tonnes. Total land based resources are estimated at 1,600 million tonnes with possibly another 700 million tonnes in deep sea nodules.

WORLD MINE PRODUCTION, 1989-90
('000 tonnes of contained copper and % of total 1990)

Developed	1989	1990	% 1990	Developing	1989	1990	% 1990	Centrally Planned	1989	1990	% 1990
Australia	295	327	(3.6)	Brazil	44	36	(0.4)	Bulgaria	39	33	(0.4)
Canada	723	802	(9.9)	Chile	1609	1588	(17.6)	China	380	360	(4.0)
Finland	14	13	(0.1)	India	53	52	(0.6)	Mongolia	135	140	(1.5)
Japan	15	13	(0.1)	Indonesia	149	170	(1.9)	Poland	385	329	(3.6)
Norway	16	20	(0.2)	Iran	68	60	(0.7)	USSR	950	900	(10.0)
Portugal	104	160	(1.9)	Mexico	249	291	(3.2)	Others	54	46	(0.5)
S Africa	197	197	(2.2)	Namibia	31	33	(0.4)				
Spain	28	15	(0.2)	Papua New							
Sweden	69	73	(0.8)	Guinea	205	170	(1.9)				
Turkey	45	44	(0.5)	Peru	364	318	(3.5)				
USA	1498	1587	(17.6)	Philippines	193	182	(2.0)				
Yugoslavia	119	119	(1.3)	Zaire	441	356	(3.9)				
Others	1	1	(-)	Zambia	510	496	(5.5)				
				Others(a)	99	95	(1.0)				
Totals	3124	3371	(37.4)		4015	3847	(42.6)		1943	1808	(20.0)
Grand Totals	1989-	9082									
	1990-	9026									

(a) Botswana 21, Malaysia 24, Morocco 15, Oman 14, Zimbabwe 15, Others 6 (all in 1990).

WORLD REFINERY PRODUCTION, 1989-90
('000 tonnes metal and % of total 1990)

Developed	1989	1990	% 1990	Developing	1989	1990-	% 1990	Centrally Planned	1989	1990	% 1990
Australia	255	274	(2.6)	Argentina	11	12	(0.1)	Albania	15	9	(0.1)
Austria	46	50	(0.5)	Brazil	167	157	(1.5)	Bulgaria	56	24	(0.2)
Belgium	344	332	(3.1)	Chile	1071	1192	(11.1)	China	450	460	(4.4)
Canada	515	516	(4.8)	Egypt	4	4	(-)	Czecho-			
Finland	56	65	(0.6)	India	42	39	(0.4)	slovakia	27	25	(0.2)
France	49	52	(0.5)	Iran	40	43	(0.4)	E Germany	94	57	(0.5)
W Germany	475	476	(4.4)	S Korea	179	187	(1.7)	Hungary	6	6	(0.1)
Italy	83	83	(0.8)	Mexico	144	152	(1.4)	N Korea	40	30	(0.3)
Japan	990	1008	(9.4)	Oman	15	12	(0.1)	Poland	390	346	(3.2)
Norway	35	36	(0.3)	Peru	224	182	(1.7)	Romania	39	29	(0.3)
Portugal	-	-	(-)	Philippines	132	126	(1.2)	USSR	1345	1260	(11.7)
S Africa	144	133	(1.2)	Taiwan	43	16	(0.1)				
Spain	166	171	(1.6)	Zaire	204	173	(1.6)				
Sweden	95	97	(0.9)	Zambia	470	479	(4.5)				
Turkey	86	84	(0.8)	Zimbabwe	24	24	(0.2)				
UK	119	121	(1.1)								
USA	1954	2017	(18.8)								
Yugoslavia	151	151	(1.4)								
Totals	**5563**	**5666**	**(52.8)**		**2770**	**2798**	**(26.0)**		**2482**	**2276**	**(21.1)**
Grand Totals	**1989 -**	**10815**									
	1990 -	**10740**									

The table includes metal refined from scrap.

COPPER

WORLD MINE AND METAL CAPACITIES
('000 tonnes of metal)

	Mine 1990	Refinery 1990
Developed		
Australia	383	324
Canada	987	640
Japan	22	1201
S Africa	209	172
USA	1864	2472
Others	506	2092
Total	**3971**	**6901**
Developing		
Chile	1735	1302
Mexico	380	190
Papua New Guinea	172	-
Peru	418	275
Philippines	241	138
Zaire	729	250
Zambia	597	605
Others	706	857
Total	**4973**	**3617**
Centrally Planned		
China	428	490
Mongolia	180	-
Poland	460	420
USSR	700	1090
Other	53	215
Total	**1821**	**2215**
TOTAL	**10765**	**12733**

Effective mine capacity falls well short of nominal capacity, especially in Africa.

SECONDARY PRODUCTION:WESTERN WORLD
('000 tonnes metal 1989-90)

	Production of Secondary Refined Copper		Direct scrap used by Manufacturers	
	1989	1990	1989	1990
European Community	604	599	859	783
Japan	231	262	649	634
USA	477	447	416	929
Others	297	293	209	785
Total	**1609**	**1601**	**3233**	**3131**

RESERVE/PRODUCTION RATIOS

Static Reserve Life (years): 36
Ratio of identified reserve base to
cumulative demand 1991-2010: 2.2 : 1 (land based only)

CONSUMPTION OF REFINED METAL

| | '000 tonnes | | %p.a. growth rates. | | |
	1989	1990	1960s	1970s	1980s
European Community	2709	2811	2.3	1.3	1.2
Japan	1447	1578	10.4	4.9	3.2
USA	2207	2152	4.3	..	1.4
Others	2297	2261	4.9	5.0	3.7
Total Western World	**8660**	**8802**	**4.3**	**2.3**	**2.2**
Total World	**11018**	**10821**	**4.4**	**2.7**	**1.5**

END USE PATTERNS (%)

	USA(a) 1990	Japan 1989	W. Europe 1989
Electrical	26	26	24
Construction	40	27	36
General engineering	14	13	17
Transport	11	15	9
Miscellaneous	9	19	14

Source: USBM & CRU
(a)Building wire etc included in construction rather than electrical use.

VALUE OF CONTAINED METAL IN ANNUAL PRODUCTION

$25 billion (refined metal at 1991 average price).

SUBSTITUTES

Vulnerable to substitutes on price grounds, technical superiority, or weight both directly (eg: aluminium in electrical uses and car radiators, optical fibres in telecommunications or plastics in plumbing), or indirectly (eg: aluminium or plastics for brass). Miniaturisation of components is also important. Not all substitution is, however, one way: copper can hold its own in many major uses.

TECHNICAL POSSIBILITIES

Possible source in deep sea nodules in the next century. The expansion of in situ leaching, and electrochemical processing methods which are both well advanced. Uses in solar energy and marine applications (ships' cladding and fish farming), and expanded markets in roofing.

COPPER

PRICES

	1986	**1987**	**1988**	**1989**	**1990**	**1991**
Cathode, higher grade/ Grade A						
LME Cash						
cents/lb	62.4	81.0	118.1	129.2	120.9	106.1
(£/tonne)	936.3	1078.1	1459.7	1734.1	1496.0	1319
LME Cash (cents/lb)						
Real Dec 1991 prices	72.4	91.3	128.9	134.5	121.3	105.7
LME Range						
£/ tonne	873-1027	869-1715	1119-2005	1470-2007	1256-1812	1149-1500

Most copper is sold through annual supply contracts but producer pricing tends to operate in protected markets such as Japan, S Korea, Taiwan and India and in major producing nations like Australia, Canada and S Africa. US producers sell partly on a list basis but one linked to Comex prices. Elsewhere, prices are linked to LME, or to a lesser extent Comex, prices which fluctuate markedly. Even in other markets the LME price exerts a major influence.

Copper prices respond rapidly to changes in demand and stocks. They can also be sensitive to world financial and political events.

MARKETING ARRANGEMENTS

Around 400 mines but far fewer companies. Around 50% of production is under state ownership or control.

CIPEC, the intergovernmental organisation, which aimed to co-ordinate measures to raise copper earnings, was largely ineffective. It has contracted, as many members have left, and its residual co-ordinating functions are now based in Chile.
Discussions are well advanced for the establishment of an International Copper Study Group under UN auspices bringing together producing and consuming countries to discuss statistics and market trends. The International Copper Association is a private organisation for technical development and market promotion. Most major producing companies are members.

REAL PRICES 1979 to 1991
Copper, LME Cash

Index Numbers 1991 = 100

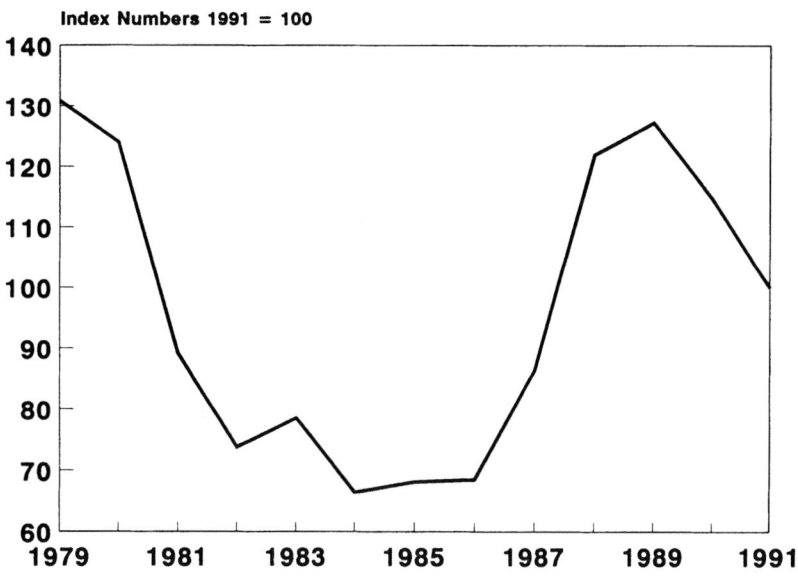

WORLD PRODUCTION 1979 to 1991
Copper, Refined Metal

Index Numbers 1991 = 100

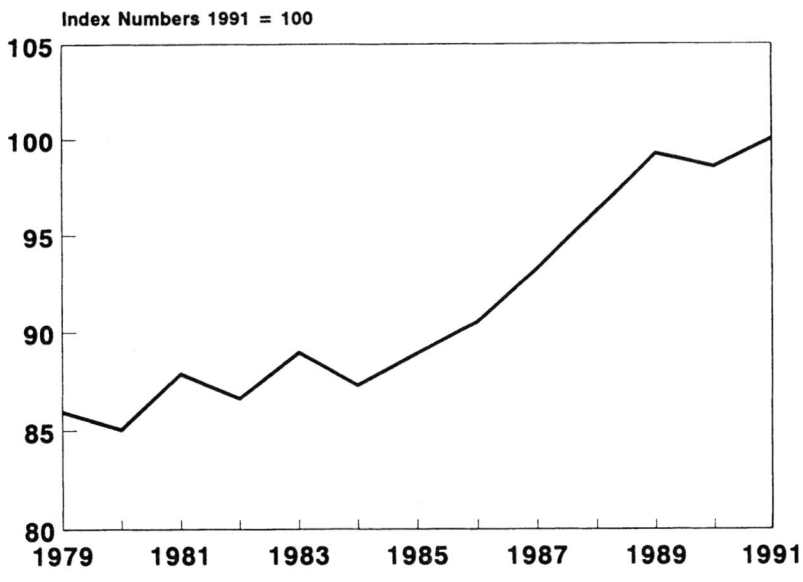

COPPER

SUPPLY AND DEMAND BY MAIN MARKET AREA

	UK 1989	UK 1990	EC(12) 1989	EC(12) 1990	Japan 1989	Japan 1990	USA 1989	USA 1990
Production								
('000 tonnes Cu content)								
Mine	0.5	0.9	131.9	174.0	14.7	13.0	1497.8	1587.2
Smelter	-	-	521.2	498.4	1005.5	1040.6	1479.5	1463.3
of which Secondary	-	-	224.3	193.6	123.2	147.5	359.1	304.8
Refined	119.0	121.6	1236.3	1263.6	989.6	1008.0	1953.7	2017.4
of which Secondary(a)	70.4	74.6	380.1	405.1	107.4	114.8	117.8	142.4
Direct scrap used by								
manufacturers	129.7	126.3	858.9	783.0	649	634	916.4	929.2

(a) i.e.fed directly to refineries.

	UK 1989	UK 1990	EC(12) 1989	EC(12) 1990	Japan 1989	Japan 1990	USA 1989	USA 1990
Net Imports								
('000 tonnes Cu content)								
Ores and								
concentrates	0.2	0.5	207	157	892.5	930.4	47.5	153.5
Blister	65.3	64.0	402	330(a)	21.9	27.4	115.0	84.2
Refined	259.9	251.0	1394	1637	487.2	618.8	339.2	287.2
Total	**325.4**	**315.5**	**2003**	**2124**	**1401.6**	**1576.6**	**501.7**	**524**

Source of Net Imports (%)	UK 1989	UK 1990	EC(12) 1989	EC(12) 1990	Japan 1989	Japan 1990	USA 1989	USA 1990
Ores and concentrates								
Norway			4	4				
Canada			10	13	26	28	2	1
USA			15		17	13		
European Community	100	100			4	5	1	17
Turkey				1				
Australia					5	7	1	
Chile			19	30	7	10	7	
Mexico			1	2	2	1	86	81
Peru			10	5	2	3	2	
Indonesia				10	9	7		
Malaysia					3	3		
Morocco			6	8				
Philippines					13	12		
Papua New Guinea			33	20	12	7		
Poland								
Others			2	7	2	4	1	1
Blister								
European Community								
Japan							22	18
Australia	2		5					
Finland	12	12	6	7				
Sweden								
S Africa	14	13	19				1	
Namibia	1						1	
Chile	26	30	24	17			36	49
Mexico				8			30	23
Peru	45	40	28	13	100	94	4	1
Zaire		2	14	37		4	5	9
Others		3	4	8		2	1	

	UK 1989	UK 1990	EC(12) 1989	EC(12) 1990	Japan 1989	Japan 1990	USA 1989	USA 1990
Refined								
Australia	14	16	4	5	5	7		
Canada	21	20	6	6	1		53	64
European Community	13	7					2	
S Africa	1	5		2	2			
Norway	3	3	2	2				
Sweden	4	10	2	2			1	
USA	2	2	2	2	10	17		
Chile	19	21	33	31	30	31	30	28
S Korea				3				
Peru	10	7	6	5	5	5	4	1
Philippines		..	..	10	9			
Zaire		8	11	1	1	3	1	
Zambia	2	3	10	9	30	25		
Brazil	1	2	1	1			6	3
China					1	1		
Poland	1	1	6	7			2	
USSR	4	5	6	10	1	1		
Others	6	2	9	9	1	1	2	
Net Exports ('000 tonnes Cu content)								
Ores and Concentrates	0.3	1.3	57.3	110.5	-	-	360.5	258.2
Blister	0.5	0.3	0.4	1.2	32.5	29.0	5.5	6.4
Refined	10.6	11.3	64.5	75.8	32.8	50.7	145.8	212.7
Total	**11.4**	**12.9**	**122.2**	**187.5**	**65.3**	**79.7**	**511.8**	**477.3**
Consumption ('000 tonnes Cu content) Refined including secondary but not direct scrap	324.7	317.2	2709.2	2810.8	1446.6	1577.5	2206.6	2152.0
Import Dependence (%) Imports as % of consumption	100	99	74	76	97	100	33	24
Imports as % of consumption and net exports	97	96	71	71	93	95	18	20
Share of World Consumption (%) Total refined:								
Western World	4	4	31	32	17	18	25	24
Total World	3	3	25	26	13	15	20	20
Consumption Growth (% p a.)								
1970s	-1.9		1.3		4.9		..	
1980s	-2.5		1.2		3.2		1.4	

FLUORSPAR

WORLD RESERVES
(million tonnes contained fluorspar and % of total)

Developed			Developing			Centrally Planned		
France	10	(4.2)	Kenya	2	(0.8)	China	27	(11.3
Italy	6	(2.5)	Mexico	19	(7.9)	Mongolia	50	(20.9
S Africa	30	(12.6)	Thailand	1	(0.4)	USSR	62	(25.9
Spain	6	(2.5)	Namibia	3	(1.3)	Others	12	(5.9
UK	2	(0.8)	Others	4	(1.7)			
USA	1	(0.4)						
Canada	2	(0.8)						
Others	2	(0.8)						
Totals	**59**	**(24.7)**		**29**	**(12.1)**		**151**	**(63.2**
Grand Total		**239**						

Pure fluorspar, CaF_2, contains 51% calcium and 49% fluoride. Three principal grades are available commercially; acid grade with 97%+ CaF_2; ceramic grade 85-96% CaF_2; and metallurgical grade 60%+ CaF_2. The above reserve figures refer to 100% CaF_2 equivalent. On the same basis, the reserve base is 340 million tonnes.

In addition fluorspar is extracted from phosphate rock. Total world reserves from this source are estimated at 330 million tonnes of fluorspar equivalent.

WORLD MINE PRODUCTION, 1989-90, and PRODUCTIVE CAPACITY, 1989
('000 tonnes gross weight and % of production 1990)

	Mine Production							
		1989			1990			
	Acid & Ceramic grade	Metal- lurgical grade	Total	Acid & Ceramic grade	Metal- lurgical grade	Total	% of Production 1990	Productive Capacity 1989
Developed								
Canada	50	-	50	25	-	25	(0.5)	75
France	159	50	209	155	50	205	(4.0)	290
W Germany	65	9	74	65	9	75	(1.5)	100
Italy	67	60	127	67	60	127	(2.5)	200
S Africa	319	47	368	269	41	310	(6.1)	680
Spain	130	5	135	130	5	135	(2.6)	320
Turkey	-	13	15	-	15	13	(0.2)	n.a
UK	50	72	122	50	69	119	(2.4)	320
USA	66	-	66	63	-	63	(1.2)	75
Others	-	3	3	-	3	3	(0.1)	n.a
Total	**906**	**261**	**1167**	**825**	**250**	**1075**	**(21.1)**	**1985**
Developing								
Argentina	9	14	23	8	12	20	(0.4)	35
Brazil	57	39	96	55	35	90	(1.8)	75
India	10	18	23	10	13	23	(0.4)	30
Kenya	115	-	95	95	-	95	(1.9)	105
Mexico	386	393	779	307	327	634	(12.4)	1220
Morocco	105	-	105	105	-	105	(2.1)	90
Thailand	-	98	98	-	100	100	(2.0)	180
Tunisia	54	-	54	53	-	53	(1.0)	45
Namibia	15	-	15	30	-	30	(0.6)	50
Others	-	32	32	-	31	31	(0.6)	n.a.
Total	**731**	**589**	**1320**	**603**	**518**	**1181**	**(23.2)**	**1830**
Centrally Planned								
China	1200	500	1700	1200	300	1500	(29.4)	2000
Czecho- slovakia	47	48	95	47	48	95	(1.8)	100
E Germany	20	70	90	10	60	70	(1.4)	100
Mongolia	-	750	750	-	750	750	(14.7)	925
N Korea	-	40	40	-	40	40	(0.8)	45
Romania	-	5	5	-	5	5	(0.1)	30
USSR	197	213	410	182	198	380	(7.5)	590
Total	**1464**	**1626**	**3090**	**1439**	**1401**	**2840**	**(55.7)**	**3790**
TOTAL	**3101**	**2476**	**5577**	**2927**	**2169**	**5096**	**(100.0)**	**7605**

Note:The split between production of acid and metallurgical grade is partly estimated.
The contained fluorspar content of total production was approximately 5 million tonnes in 1989 and 4.6 million tonnes in 1990.

FLUORSPAR

RESERVE/PRODUCTION RATIOS

Static Reserve Life (years)   Fluorine in fluorspar: 52
Ratio of identified reserve base to
cumulative demand 1991-2010:   Fluorine in fluorspar: 4.4:1

CONSUMPTION

	Averages '000 tonnes		% p.a. growth rates	
	1989	1990	1970s	1980s
European Community	930	950	-0.4	-0.6
Japan	709	567	-0.5	3.4
United States (apparent)	693	587	-2.5	-4.4

Reported US consumption is 642,000 tonnes in 1989 and 564,000 tonnes in 1990.

END USE PATTERNS (%)

	USA 1990	Japan 1989
Steel production	20*	33
Primary aluminium production }	78	..
Chemicals }		30
Glass, enamel and other uses	2	37

*Higher proportion in many other countries.

VALUE OF ANNUAL PRODUCTION

$710 million (at average 1991 prices).

SUBSTITUTES

Some substitution possible in steelmaking but rarely totally satisfactory.

Gaseous hydrocarbons and carbon dioxide are increasingly used in aerosol propellants. The Montreal Convention envisages reduced usage of ozone-depleting CFCs, and pressures to eliminate them completely are gathering rapid momentum. The latest objective is to phase them out by 2000-2010, but substitution is proceeding more rapidly.

TECHNICAL POSSIBILITIES

Conservation in the steel industry, recycling, changing technology in the aluminium industry and environmental concern over fluorocarbons in propellants are reducing demand. Developments in industrial and medical applications will only partly offset these reductions.

Further exploitation of phosphate rock as source of fluorspar in the USA.

PRICES

	1986	1987	1988	1989	1990	1991
Mexican fob Tampico Metallurgical $/tonne	74.5	52.1	62.8	79.1	90.3	90.9
Real Dec 1991 prices	86.5	58.9	68.5	82.2	90.6	90.6
USA Illinois district bulk $/short ton Acid spar	170.5	170.5	170.5	170.5	183.4	192.5
Real Dec 1991 prices	197.6	192.5	185.9	177.3	183.8	191.8

Source:Industrial Minerals

Mainly producer pricing.

MARKETING ARRANGEMENTS

Although there is a large number of small firms participating in fluorspar mining, world production is dominated by large companies.

REAL PRICES 1979 to 1991
Fluorspar, Acid spar Illinois district

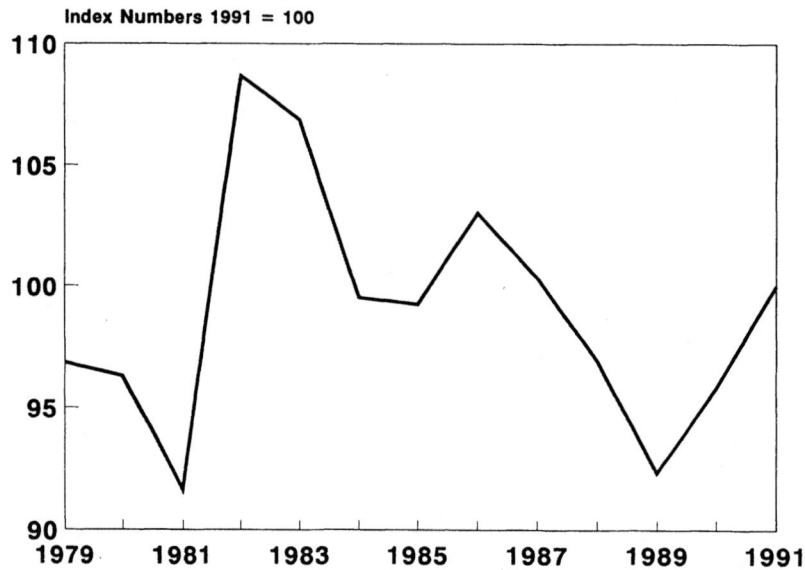

WORLD PRODUCTION 1979 to 1991
Fluorspar

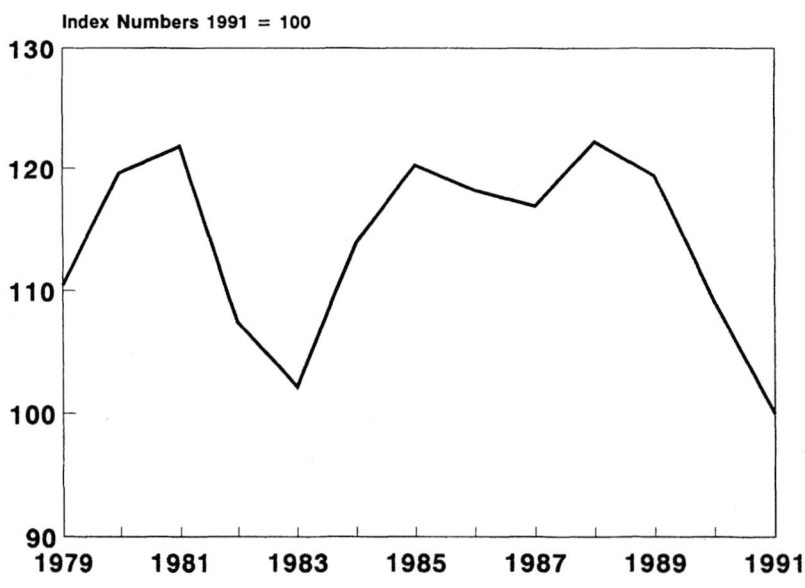

SUPPLY AND DEMAND IN MAIN MARKET AREA

	UK		EC(12)		Japan		USA	
	1989	1990	1989	1990	1989	1990	1989	1990
Production								
('000 tonnes)								
Gross	122	118.5	667	661	-	-	66	64
Fluorspar equivalent from phosphate rock	-	-	-	-	-	-	91	82
Net Imports								
('000 tonnes)								
Acid grade more than 97% CaF_2	13.8	13	148.9	155.8	323	237	554	406
Met grade less than 97% CaF_2	6.3	4.5	171.9	185.6	386	330	102	107
Fluorspar equivalent from hydrofluoric acid & cryolite							127	109
Source of Net Imports (%)								
Acid Grade								
Brazil								1
Canada							7	5
European Community								2
S Africa			62	41	19	15	21	21
China			19	32	78	72	29	31
Namibia								1
Kenya				15		13	1	2
Mexico					3		33	35
Morocco			11	9			7	3
Others and undefined		n.a.	n.a.	8	3			1
Other Grades								
European Community	68	100						
S Africa			25	25				8
China			40	54	80	86	3	26
Kenya			11	7				
Mexico	32		7	7	4		97	66
Morocco			8	4				
Tunisia			7					
Thailand					14	14		
Others and undefined	-	..	2	3	2			

FLUORSPAR

	UK		EC(12)		Japan		USA	
	1989	1990	1989	1990	1989	1990	1989	199
Net Exports ('000 tonnes)								
Acid grade + 97% CaF_2	8.4	4.0	41.2	27.8	0.2	0.1	2.8	9.
Other grades - 97% CaF_2	-	-	16.5	24.9	0.3	0.2	2.3	5
Total								
Consumption ('000 tonnes)								
Acid grade							397	32
Other grades						245	240	
All forms (inc. hydrofluoric acid, etc)	134	132 (apparent)	930	950 (apparent)	709	567 (apparent)	642	56 (reported)
							693	58 (apparen
Import Dependence								
Imports as % of consumption	15	13	34	36	100	100	95	8
Imports as % of consumption and net exports	14	13	32	34	100	100	94	8
Share of World Consumption (%)								
Total World	2	3	17	19	13	11	12	1
Consumption Growth (% p.a.)								
1970s	0.2		-0.4		-0.5		-2.5	
1980s	-1.2		-0.6		3.4		-4.4	

GALLIUM

WORLD RESERVES

Most gallium is recovered as a by-product of the extraction of alumina from bauxite, with recovery from the smelting of zinc ores as the second major source. Gallium's nature as a by-product, and the protective patents covering the recovery process prevent any precise measurement of reserves. Nonetheless the world's bauxite reserves are estimated to contain over 100,000 tonnes of gallium, and zinc resources 6,500 tonnes. Only a small percentage though is economically recoverable.

WORLD PRODUCTION & CAPACITY

Estimated primary production was about 60 tonnes in 1989 and 55 tonnes in 1990 (including USSR production). Recycled gallium makes up over one-third of total supply. Detailed production data are not available but capacities were as follows at the end of 1990.

Tonnes

Developed		Centrally Planned	
USA	12(a)	Czechoslovakia	3
France	20	Hungary	4
Germany	15	USSR	30
Norway	5(a)	China	8
Japan	17		
Australia	50		
Total	**119**		**45**
Grand Total	**164**		

(a) On standby.

Much of this capacity was on standby with less than half operational, often below capacity, at end 1990. The Australian plant opened in April 1989 and closed in late 1990. It shipped crude product to France for refining. Germany, France and Japan were the main producers in 1989 with Australia replacing France in 1990.

RESERVE/PRODUCTION RATIOS
Very large because of substantial resources of bauxite and zinc. Any supply bottleneck would be caused by the available processing capability.

CONSUMPTION

	(kilograms)		% p.a. growth rates	
	1989	1990	1970s	1980s
European Community	n.a	n.a	n.a	n.a
Japan(a)	54000	60000	n.a	16.6
United States	9667	9860	21.4	1.1

(a) Including scrap.

GALLIUM

END USE PATTERNS, 1990 (USA) (%)

Optoelectronic devices	60
Integrated circuits	17
Research and development	9
Others	14

VALUE OF CONTAINED METAL IN ANNUAL PRODUCTION

$21 million approximately for primary metal(at average 1991 prices).

SUBSTITUTES

Liquid crystals made from organic compounds are used in visual display panels as substitutes for light emitting diodes. Silicon and germanium compete with gallium in many semiconductor applications.

There are no effective substitutes in some defence uses.

TECHNICAL POSSIBILITIES

Gallium could be recovered from coal ash and coal, and extracted from polymetallic ores by leaching.

Increasing use in gallium based electronic devices and in equipment converting solar energy to electricity.

PRICES

	1986	1987	1988	1989	1990	1991
US Metal $/kg(a)	525	525	525	525	525	525
US Metal Real Dec 1991 prices	609.5	592.8	573.3	546.1	525.5	522.9

(a) nominal

Prices are listed by producers, but discounting is common.

MARKETING ARRANGEMENTS

Only a handful of companies extracts gallium in the main metals processing countries.

REAL PRICES 1979 to 1991
Gallium, US metal

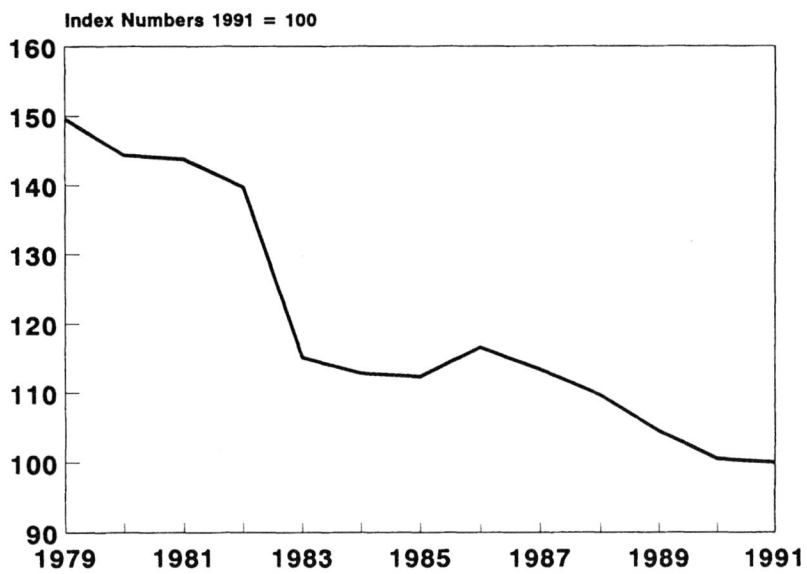

Index Numbers 1991 = 100

WORLD PRODUCTION 1979 to 1991
Gallium

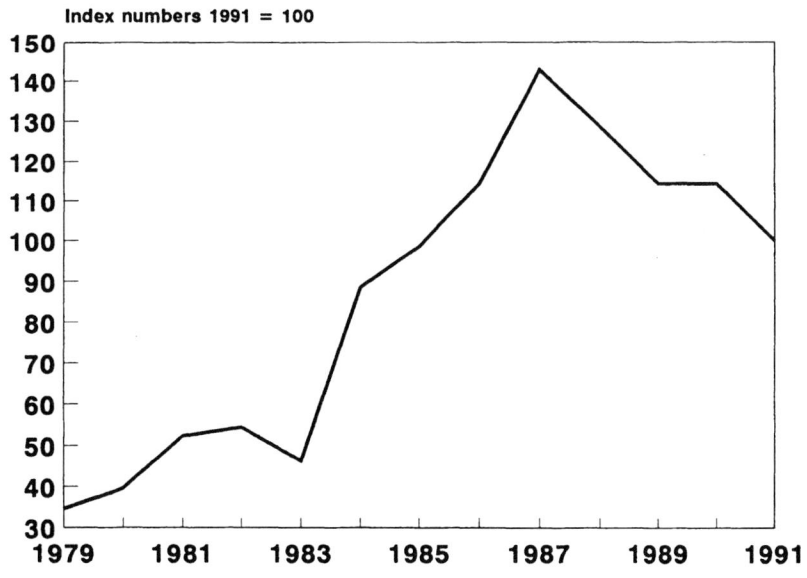

Index numbers 1991 = 100

GALLIUM

SUPPLY AND DEMAND BY MAIN MARKET AREA

	UK		EC(12)		Japan		USA	
	1989	1990	1989	1990	1989	1990	1989	1990
Production (kg)								
Primary	-	-	n.a	n.a	32000	34000	n.a	
Net Imports (kg)	29000(a)	33000(a)	27000(a)	28000(a)	16000+	20314+	18284	9894
(a)Gallium, thallium and indium.								
Source of Net Imports (%)								
Canada	14		15					
China			7					4
European Community	62	52					85	79
Japan		..		7				3
Switzerland			15				14	
USA	7	3	19	25				
Hungary								3
USSR	7	6	7	14				
Others and unidentified	10	39	37	54			1	11
Net Exports (kg)	13000(a)	65000(a)	80000(a)	119000(a)	11956(a)	21684(a)	n.a	n.a
(a)Gallium, thallium and indium.								
Consumption (kg)	n.a	n.a	n.a	n.a	54000	60000 recycled	9667 recycled	9860
Import Dependence (%)								
Imports as % of consumption	n.a	n.a	n.a	n.a	n.a	n.a	100	100
Imports as % of consumption and net exports	n.a	n.a	n.a	n.a	n.a	n.a	n.a	n/a
Share of World Consumption (%)								
Total World	n.a	n.a	n.a	n.a	c.63	c.65	c.11	c.11
Consumption Growth (% p.a.)								
1970s	n.a		n.a		n.a		21.4	
1980s	n.a		n.a		16.6		1.1	

GERMANIUM

WORLD RESERVES

Germanium is obtained as a by-product of zinc or copper-zinc ores. No reliable data are available for the reserves of large tracts of the world. The US Bureau of Mines estimates the combined reserves of Canada, the United States, Europe and Africa at 2,150 tonnes, with substantial reserves also available in Centrally Planned economies. US reserves are estimated at 450 tonnes, and Zaire's at 200 tonnes, within the overall total.

Very large potential resources are contained in certain coals, and germanium might be recovered from ash and flue dusts.

WORLD PRODUCTION & CAPACITY

Because of its by-product nature no data are available for mine production of germanium. Data on refinery output are scarce, but productive capacity was estimated as follows for end 1990.

Developed		Centrally Planned	
Canada	10	China	10
USA	60	USSR & Europe	40
Japan	35		
Belgium	50		
Other Europe(a)	65		
Total	**220**		**50**
World Total	**270**		

(a) Belgium (MHO), France (Metaleurop), Germany, Italy and Austria. This capacity includes both operating plants and those on standby which can be reopened at minimal cost.

World Refinery Production was about 88 tonnes in 1989 and 52 tonnes in 1990.

RESERVES/PRODUCTION RATIOS

Static Reserve Life (years):	large
Ratio of identified reserves to cumulative demand 1991-2010:	large

CONSUMPTION

	'000 tonnes		% p.a. growth rates	
	1989	1990	1970s	1980s
European Community	c.35	c.35	n.a	n.a
Japan	10.3(a)	9.1(a)	2.1	-5.2
United States	36	34	4.0	1.9

(a)Metal and oxide.

GERMANIUM

END USE PATTERNS, 1991 (USA) (%)

Infra-red systems	62
Fibre optics	14
Semiconductors	7
Detectors	7
Others	10

VALUE OF CONTAINED METAL

$50 million (at average 1991 free market prices).

SUBSTITUTES

Silicon has replaced germanium in some electronic applications but not in high-frequency or high-power applications. In infra-red guidance systems zinc selenide or germanium glass can substitute for germanium metal but at the expense of performance.

TECHNICAL POSSIBILITIES

Substitute materials could become available for use in fibre optics. Recovery from coal ash and flue dusts.

PRICES

	1986	1987	1988	1989	1990	1991
Germanium Dioxide Electronic grade Producer price fob Paris airport						
ECU/kg	560	560	560	412.7	400	400
Metal, Zone refined 50 ohm-cm. Producer price fob Paris airport						
ECU/kg	925	925	925	679	660	660
$/kg	908	1065.8	1091.4	748.8	842.2	822.6
Real Dec 1991 prices	1054.7	1206.6	1192	776.9	843	819.3
Metal, refined US free market refined 50 ohm-cm $/kg						
Range	-	-	-	556-629	380-550	340-410
Average	-	-	-	559	461	400

Source: Metal Bulletin

Germanium is a by-product of zinc, and certain copper-zinc ores, extracted in refining. It is mainly producer priced with a small dealer market.

MARKETING ARRANGEMENTS

Belgium refines germanium from Zairois ores. There are relatively few producers and consumers. Commercial availability is governed by the rate at which germanium-bearing materials are processed and refined. There is some speculative activity. The collapse of the USSR contributed to oversupply in 1991. The United States established a National Defence Stockpile goal of 146 tonnes in mid 1987. Purchases to achieve this goal have been made in recent years (26.5 tonnes in 1989 and 20 tonnes in 1990).

·REAL PRICES 1979 to 1991
Germanium, Producer price, zone refined

Index Numbers 1991 = 100

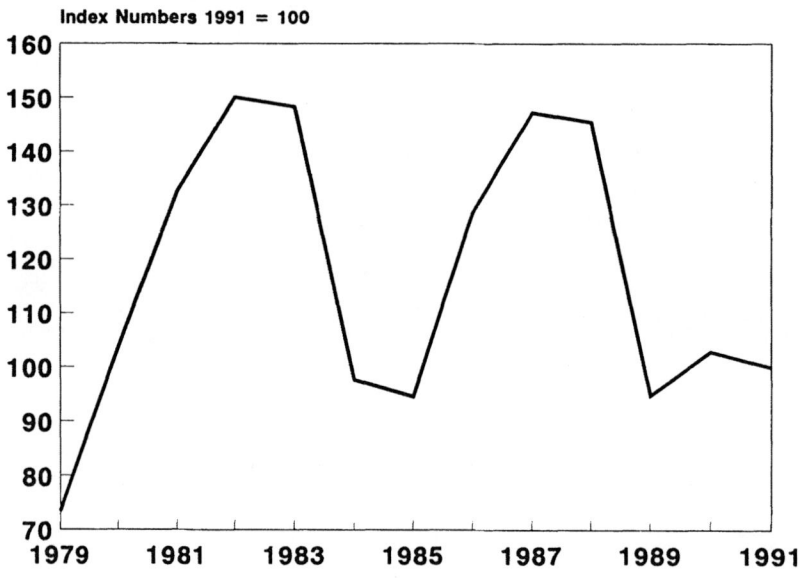

WORLD PRODUCTION 1979 to 1991
Germanium

Index Numbers 1991 = 100

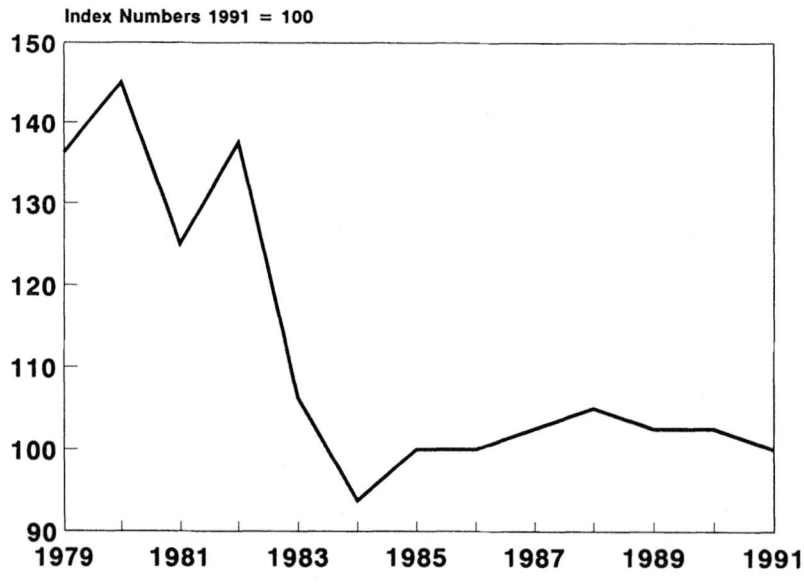

SUPPLY AND DEMAND BY MAIN MARKET AREA

	UK 1989	UK 1990	EC(12) 1989	EC(12) 1990	Japan 1989	Japan 1990	USA 1989	USA 1990
Production (tonnes)								
Mine	-	-	-	-	-	-	n.a	n.a
Refinery	-	-	c.35	c.35	13.3 oxide	12.4 oxide	20	18
							3.3 metal	3.4 metal
Net Imports (tonnes)								
Metal (incl. unwrought)	8	11	5(a)	1a)	1.1	1.7	40	50
Oxides	2	28					(incl. oxides)	

(a)Excl. Belgium-Luxembourg.

Source of Net Imports (%)								
European Community	50	100			64	14	78	77
USA	25		40	100	5	13		
China					20	73	8	4
USSR					9			1
Canada							6	11
Hong Kong								6
Others	25		60		2		8	1

Net Exports (tonnes)(incl.unwrought)	8	6	19(a)	30(a)	0.5	0.1	n.a	n.a

(a) Excl. Belgium-Luxembourg, which exported 16 to USA in 1989 and 7.5 in 1990.

Consumption (tonnes)	2	5 (apparent)	c.35(a)	c.35(a) (apparent)	10.3	9.1 (incl. oxide)	36	34

(a) Partly for germanium compounds for export.

Import Dependence								
Imports as % of consumption	100	100	n.a (near 100 based on raw materials)	n.a	11	19	n.a	n.a
Imports as % of consumption and net exports	100	100	n.a	n.a	10	18	n.a	n.a

Share of World Consumption (%)								
Total World	n.a	n.a	n.a	n.a	n.a	n.a	n.a	n.a

Consumption Growth (% p.a.)								
1970s	n.a		n.a		2.1		4.0	
1980s	n.a		n.a		-5.2		1.9	

GOLD

WORLD RESERVES

The considerable exploration activity of the past decade, and strong movements in costs relative to prices have moved ahead of detailed estimates of world gold reserves. The figures in the table are therefore highly approximate.

(tonnes of metal and % of total)

Developed			Developing			Centrally Planned		
Australia	1400	(3.2)	Brazil	940	(2.2)	USSR	6220	(14.5
Canada	1780	(4.1)	Others	6820	(15.8)	Others	n.a	
S Africa	20000	(46.5)						
USA	4770	(11.1)						
Others	1100	(2.6)						
Totals	29050	(67.5)		7760	(18.0)		6220	(14.5
Grand Total		43030						

The estimated reserve base is 49400 tonnes, excluding China and other centrally planned economies. Total world resources are estimated at 75,000 tonnes. These figures should be treated with caution as exploitation of gold deposits is heavily price-dependent. In addition above ground stocks of previously mined gold, held by both central banks and privately, are substantial. Approximately 38,000 tonnes are held officially by central banks as reserves, and 50,000 tonnes are held in the form of coin, bullion and jewellery.

WORLD MINE PRODUCTION, 1989-90
(tonnes of metal and % of total 1990)

Developed	1989	1990	% 1990	Developing	1989	1990	% 1990	Centrally Planned	1989	1990	% 1990
Australia	203.6	242.2	(11.8)	Bolivia	3.6	6.0	(0.3)	China	90.0	100.0	(4.1)
Canada	159.5	165.0	(8.0)	Brazil	100.0	80.0	(3.9)	N Korea	5.0	5.0	(0.2)
Japan	6.1	7.3	(0.4)	Chile	22.6	29.6	(1.3)	USSR	285.0	250.0	(12.2)
S Africa	607.1	602.8	(29.5)	Colombia	27.1	28.0	(1.4)T	Others	1.7	1.5	(0.1)
Spain	8.2	8.4	(0.4)	Dominican R	5.2	4.3	(0.3)				
USA	265.7	290.2	(14.6)	Ghana	13.3	16.8	(0.8)				
Others	20.4	20.9	(1.0)	Mexico	8.6	8.8	(0.4)				
				P New Guinea	27.5	31.0	(1.5)				
				Peru	10.6	8.50	(0.4)				
				Philippines	35.3	35.0	(1.7)				
				Venezuela	3.9	7.7	(0.4)				
				Zaire	2.0	2.6	(0.1)				
				Zimbabwe	16.0	16.9	(0.8)				
				Others	69.2	82.6	(4.0)				
Totals	1270.6	1336.8	(65.3)		344.9	355.3	(17.3)T		381.7	356.5	(17.4)
Grand Totals	1989 -	1997.2									
	1990 -	2048.6									

PRODUCTIVE CAPACITY, 1990 (Major Producers)
(tonnes of metal)

Developed		Developing		Centrally Planned	
Australia (a)	250	Brazil	95	USSR	285
Canada	170			China	120
S Africa	630				
USA	300				
Totals	**1350**		**95**		**405**

(a) Includes Papua New Guinea

The combined capacity of all other gold-producing nations amounts to 420 tonnes, giving a total world gold mine production capacity of 2270 tonnes.

RESERVE/PRODUCTION RATIOS

Based on demand for fabricated gold — i.e. excluding monetary and 'investment' uses. World bullion stocks are ignored. These will make up any shortfalls between mined output and demand.

Static Reserve Life (years):	22
Ratio of identified reserve base to cumulative demand 1991-2010:	0.7 : 1

OVERALL BALANCES OF SUPPLY AND DEMAND IN THE WESTERN WORLD

(tonnes)	1988	1989	1990
Mine Production	1547	1677	1744
Net Trade with Centrally Planned Economies	263	266	425
Net Official Sales	-	217	-
Net Official Purchases	265	-	66
Scrap	351	360	490
Available Supplies	1876	2520	2593
Fabricated gold in Developed Countries	999	1148	1214
Fabricated gold in Less Developed Countries	924	1165	1225
Bullion Holdings (excluding Europe and N America)	461	514	234
Net Implied Investment (disinvestment) in Europe and N America	(508)	(307)	(80)

Source: Gold Fields Mineral Services

GOLD

INDUSTRIAL USAGE OF GOLD IN THE WESTERN WORLD (excludes coins & medals)

	tonnes		% p.a. growth rates	
	1989	1990	1970s	1980s
Jewellery	1907(a)	2032(a)	-4.1(b)	14.8(a)
Electronics	137	146	0.6	4.4
Dentistry	51	53	4.5	-1.9
Other industrial and decorative uses	64	64	2.2	-0.2
Total	**2159**	**2295**	**-4.2**	**4.7**
of which:				
European Community	537	609	-1.0	8.8
Japan	200	204	2.2	10.9
United States	211	201	-2.6	3.8
Other countries	1212	1281	-8.1	17.3

(a) Including scrap
(b) Excluding scrap

Source: Gold Fields Mineral Services Ltd

END USE PATTERNS, 1990 (%)

	USA	Japan	EEC	Other Western Countries
Jewellery	58	53	88	90
Electronics	19	33	4	1
Dentistry	5	7	3	1
Other industrial and decorative uses	10	7	3	1
Coins and small items for investment	8	..	2	7

Source: Gold Fields Mineral Services Ltd

VALUE OF CONTAINED METAL IN ANNUAL PRODUCTION

$24 billion (at average 1991 prices).

SUBSTITUTES

Platinum and palladium substitute to some extent but use is influenced by price relationships and by established consumer preference for gold. Silver can substitute but is more subject to corrosion. Gold-plated palladium and bright tin-nickel can be used in electronics. Titanium- and chromium-base alloys can be used in dental work.

High prices in 1979-1981 encouraged substitutes, particularly base metals clad with gold alloy in electronics/electrical industry and in jewellery products. No metal or alloy substitute has all gold's properties, and emphasis is on reduction of gold content rather than substitution.

TECHNICAL POSSIBILITIES

New gold dissolution methods and better media for solvent or resin extraction could improve production technology and utilisation of lower grade sources.

PRICES

	1986	1987	1988	1989	1990	1991
London fixing am $/troy oz	367.8	447.1	437.1	381.7	383.7	362.3
Real Dec 1991 prices	427.4	505.3	477.6	397.1	383.3	360.9

Above ground stocks of gold are very high and the willingness to add to or release from these stocks largely determines the state of the market.

MARKETING ARRANGEMENTS

S Africa and former USSR produce about half the world's output. The state of the Soviet economy tended to dictate its sales and IMF auctions plus selling from Central Bank stockpiles have in the past supplemented supply. Speculative activity, particularly in response to political tension, has transformed the market in very short time. Demand for investment related to inflationary expectations, the level of real interest rates, and exchange rates. Prices are increasingly dominated by producers' forward selling and options trading, fuelled by the willingness of Central Banks to lend into the market. Trade in gold derivatives has, to an extent, replaced the use of gold itself as a political and economic hedge.

The collapse of the USSR, and revelation about the inadequacies and uncertainties of the CIS Republics, gold production and holdings have created new issues for the gold market.

REAL PRICES 1979 to 1991
Gold, London Fixing am

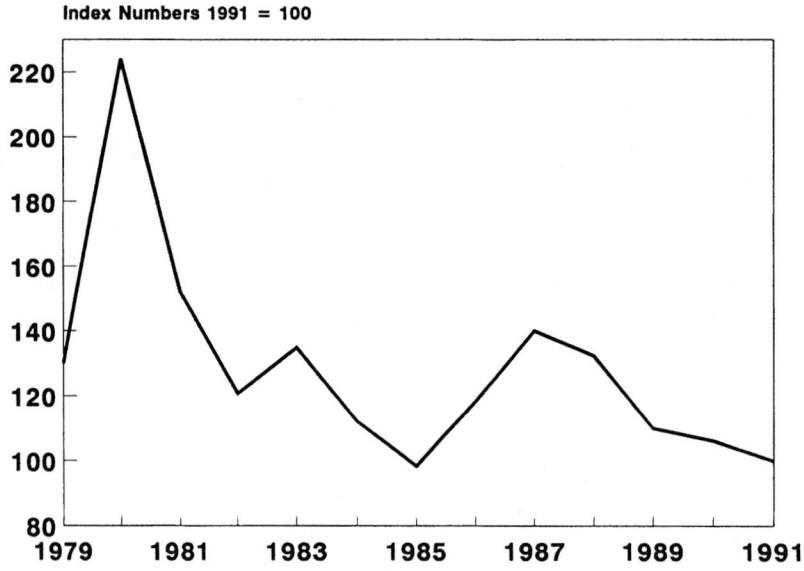

WORLD PRODUCTION 1979 to 1991
Gold

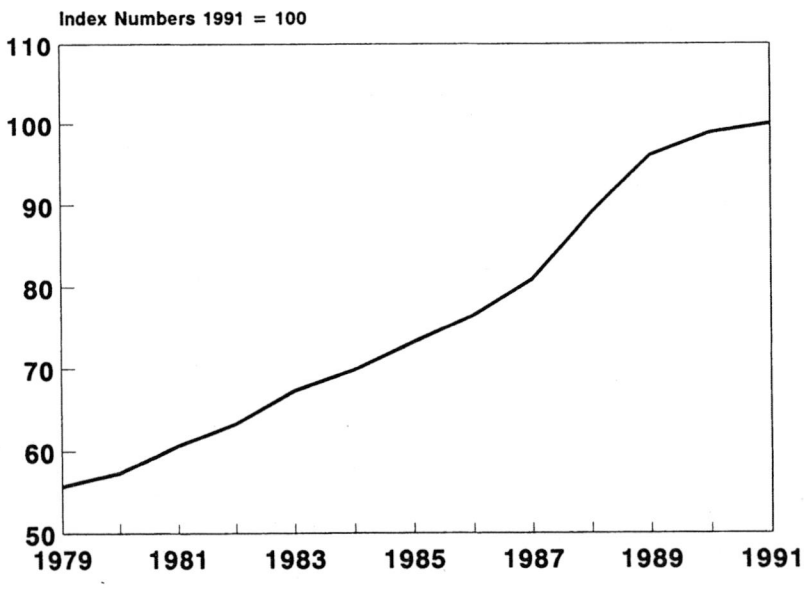

SUPPLY AND DEMAND BY MAIN MARKET AREA

	UK		EC(12)		Japan		USA	
	1989	1990	1989	1990	1989	1990	1989	1990
Production (tonnes)								
Mine production	-	-	10.3	12.1	6.1	7.3	265.5	294.2
Metal (inc. scrap)	n.a	n.a	n.a	n.a	110.3	108.2	220.8	236.9
Net Imports (tonnes)								
Ores and concentrates	n.a	n.a	n.a	n.a	n.a	n.a	2.11	1.85
Unwrought metal (inc. semi-manufactures)	569	2245	610	748	284	303	112.0	71.5
Waste and scrap	282	145	526	349	2.3	1.9	40.0	23.2

Data exclude gold imported in foreign coins.

Source of Net Imports (%)

Unwrought/wrought waste & scrap(a)	1989	1990	1989	1990	1989	1990	1989	1990
Australia				1	18	23	..	
Canada	7		6	4	10	16	49	39
European Community	37	86			11	13	5	2
S Africa	8		14	10	3	1		
Sweden			3	6				
Switzerland	2		10	17	46	30	12	24
United States		16	3	20	21	3	2	
USSR			2	2	6	6		
Poland			5					
Czechoslovakia			3	4				
Austria		1						
Bolivia	2	1	3	3			1	7
Chile	10	2	10	8			5	8
Dominican Republic							4	8
Mexico	1		1				6	1
Hong Kong			1				1	
Peru								2
Philippines							2	1
Singapore								6
Uruguay	3	1	2	2				1
China							9	
Namibia		3		6				
Sierra Leone		2		2				
Kenya	7		5					
Others (including secret)	4	4	16	16	3	6	3	5

(a) UK figures exclude unwrought refined bullion transactions.

GOLD

	UK 1989	UK 1990	EC(12) 1989	EC(12) 1990	Japan 1989	Japan 1990	USA 1989	USA 1990
Net Exports (tonnes)								
Unwrought metal	302	175	168	67	6.8	7.8	173.0	231.0
Wrought metal	86	44	167	19	11.4	23.2	10.4	6.9
Waste and scrap	156	248	559	699	-	0.3	44.5	149.5
Consumption (tonnes) Gold Fields Mineral Services Ltd	37	39 (incl. Ireland)	537	609	200	204	211	201

Import Dependence

Imports as % of consumption }
Imports as % of consumption }   Because of gold's monetary role, its use
and net exports }   as an investment medium, and the small
}   share of newly mined output in total
}   supply, import shares mean very little.

Share of World Consumption (%) (based on Gold Fields Mineral Services Ltd figures) Western World (Industrial Uses)	2	2	25	27	9	9	10	9
Consumption Growth (% p.a.) Gold Mineral Mineral Services figures								
1970s	-1.0		0.2		2.2		-2.6	
1980s	6.0		8.8		10.9		3.8	

INDIUM

WORLD RESERVES
(tonnes of contained indium and % of total)

Developed			Developing			Centrally Planned		
Australia	185	(8.49)	Bolivia	37	(1.7)	China	200	(9.1)
Canada	600	(27.2)	Mexico	60	(2.7)	E Germany	20	(0.8)
France	120	(5.4)	Peru	100	(4.5)	USSR	200	(9.1)
W Germany	50	(2.3)	Other Asia	68	(3.1)			
Japan	70	(3.2)	Africa	190	(8.6)			
USA	300	(13.5)	Other America	9	(0.4)			
Totals	1325	(60)		464	(21.0)		420	(19.0)
Grand Total	2209							

Indium is recovered principally as a by-product of processing zinc ores but it is also present in some copper, lead and tungsten ores. The reserve base is 3,600 tonnes.

WORLD REFINERY PRODUCTION, 1989-90, and PRODUCTIVE CAPACITY, 1990
(tonnes and % of Total 1990)

	Refinery Production		% of Production	Production Capacity
	1989	1990	1990	
Developed				
Canada	10	15	11	60
United States	15	15	11	30
Belgium	18	18	14	25
France	15	18	14	20
Germany	2	2	2	10
Italy	12	9	7	12
Netherlands	1	1	..	3
United Kingdom	5	6	5	6
Japan	31	27	20	50
Total	**109**	**111**	**83**	**216**
Developing				
Peru	2	4	3	6
Total	**2**	**4**	**3**	**6**
Centrally Planned				
China	13	13	10	15
USSR	6	5	4	15
Total	**19**	**18**	**14**	**30**
TOTAL	**130**	**133**	**100**	**252**

Indium is mostly recovered from dusts at lead and zinc smelters and from the purification of zinc sulphate.
In addition to the countries listed, Mexico and North Korea may have refined smaller tonnages.
Indium bearing concentrates are mined in Australia, Canada, Sweden, Ireland, USA, Peru, China and the USSR.
Some plants produce crude material that is further treated elsewhere. This means some possible double counting in the data on capacity.

INDIUM

RESERVE/PRODUCTION RATIOS

Static Reserve Life (years): 17
Ratio of reserve base to
cumulative demand 1991-2010: 0.6 : 1

CONSUMPTION

| | tonnes | | % p.a. growth rates | |
	1989	1990	1970s	1980s
European Community	c.20	c.20	n.a	n.a
Japan	c.60	c.60	n.a	c.22
United States	28	30	1.7	4.0

VALUE OF ANNUAL PRODUCTION

$39 million (at average 1991 prices).

END USE PATTERNS (%)

USA (1990)

Electrical and electronic components	15
Solders, alloys and coatings	75
Research and other uses	10

Western World (1989)

Plating	6
Alloys	25
Semiconductors	8
Indium tin oxides	45
Other uses	16

SUBSTITUTES

Substitutes exist for indium in most end uses. Silicon has tended to replace germanium-indium in transistors. Gallium can substitute in some alloys, although at greater cost, and boron carbide and hafnium can be used in nuclear reactor control rods.

Relative cost is important in determining whether substitution occurs.

TECHNICAL POSSIBILITIES

Uses in solar cells, semiconductors and indium-tin coatings for flat glass. Potential large scale use in lasers for telecommunications and consumer electronics products.

PRICES

	1986	1987	1988	1989	1990	1991
US Producer (Indium Corp) US$/kg	85.5	232.9	317.6	281.3	228.8	217.5
Real Dec 1991 prices	99.2	261.9	346.7	292.7	229.7	216.7

Supply is relatively independent of demand in that it depends on the output of zinc. A producer price coexists with a dealer market.

MARKETING ARRANGEMENTS

The world's refining facilities are limited as only a few zinc smelters recover indium as a by-product.

Output depends on the type of zinc ore processed.

REAL PRICES 1979 to 1991
Indium, US Producer

Index Numbers 1991 = 100

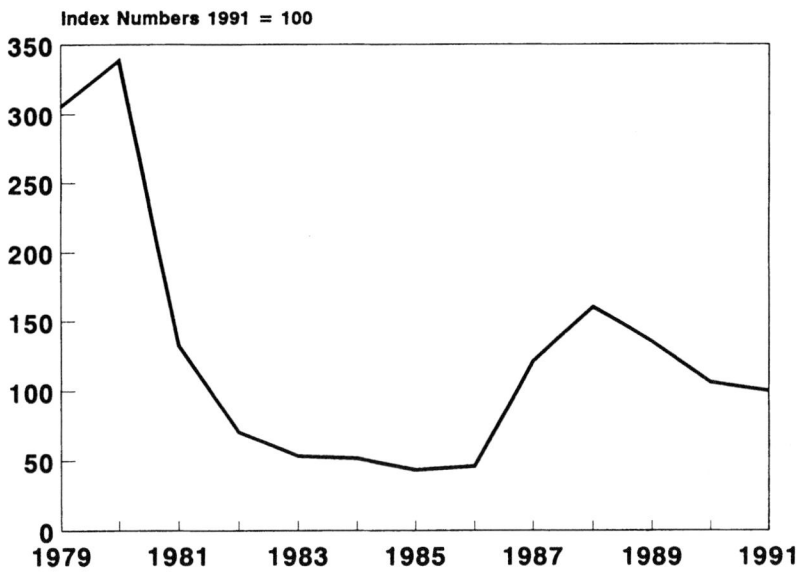

WORLD PRODUCTION 1979 to 1991
Indium

Index Numbers 1991 = 100

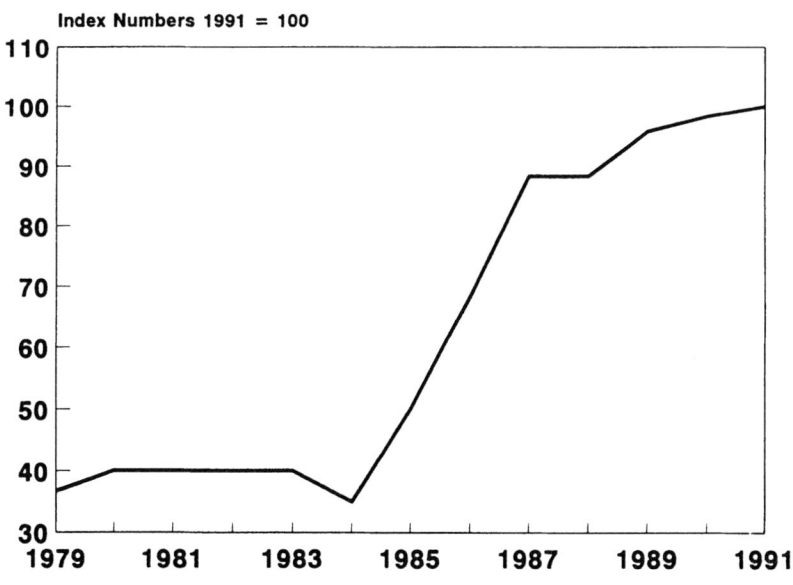

INDIUM

SUPPLY AND DEMAND BY MAIN MARKET AREA

	UK 1989	UK 1990	EC(12) 1989	EC(12) 1990	Japan 1989	Japan 1990	USA 1989	USA 1990
Production (tonnes)	c.5	c.6	c.53	c.54	49.5 (incl. treatment of crude imports) 31 (net output)	48.1 27	c.15	c.15
Net Imports (tonnes)								
Unwrought waste and scrap	incl. with gallium		incl. with gallium		31.5	36.0	26.8	30.2
Source of Net Imports (%)								
Canada					15	3	1	11
European Community					58	61	82	70
USA					12	5		
Peru							3	6
China					15	29	9	11
Others						2	1	1
USSR							2	
Japan						2	1	
Net Exports (tonnes)	incl.with gallium		incl. with gallium		incl.with gallium		n.a	n.a.
Consumption (tonnes)	n.a	n.a	n.a	n.a	c.60	c.60	28	30
Import Dependence (%)								
Imports as % of consumption	n.a	n.a	n.a	n.a	52 (a)	60(a)	54(a)	50(a)
Imports as % of consumption and net exports	n.a	n.a	n.a	n.a	53 (a)	60 (a)	54(a)	50(a)

(a) Based on production and consumption.

Share of World Consumption (%)								
Total World	n.a	n.a	c.15	c.15	c.46	c.45	c.22	c.23

Consumption growth (% p.a.)								
1970s	n.a		n.a		n.a			1.7
1980s	n.a		n.a		c.22			4.0

INDUSTRIAL DIAMONDS

WORLD RESERVES
(million carats and % of total)

Developed			Developing			Centrally Planned		
Australia	500	(51.0)	Botswana	125	(12.8)	China	10	(1.0)
S Africa	70	(7.1)	Brazil	5	(0.5)	USSR	40	(4.1)
			Zaire	150	(15.3)			
			Others	80	(8.2)			
Totals	**570**	**(58.1)**		**360**	**(36.8)**		**50**	**(5.1)**
Grand Total		**980**						

Approximately 40% of these reserves are in the form of crushing bort with the balance industrial stones. The world reserve base is 1900 million carats mainly in Australia, Botswana, S Africa, Zaire and the USSR. Synthetic industrial diamonds supplement reserves.

INDUSTRIAL DIAMONDS

WORLD MINE PRODUCTION, 1989-90,and PRODUCTIVE CAPACITY, 1990
('000 carats and % of total 1990)

	Mine Production 1989	Mine Production 1990	% of Production 1990	Productive Capacity 1990
Developed				
Australia	17483	17289	31.4	34614
S Africa	3415	3233	5.9	8645
Total	**20898**	**20522**	**37.2**	**43259**
Developing				
Angola	136	127	0.2	1270
Botswana	4869	5669	10.3	17351
Brazil	197	200	0.4	2000
Central African Rep.	90	83	0.2	415
Ghana	193	412	0.7	615
Guinea	50	49	0.1	327
Guyana	4	6	...	19
India	1.5	1.5	...	15
Indonesia	0.6	0.6	...	12
Ivory Coast	37	37	0.1	375
Swaziland	27	21	...	42
Lesotho	2	2	...	6
Liberia	52	52	0.1	150
Namibia	44	37	0.1	748
Sierra Leone	30	29	0.1	298
Tanzania	3	1	...	12
Venezuela	150	156	0.3	625
Zaire	16386	17602	31.9	24400
Total	**22272**	**24485**	**44.4**	**48680**
Centrally Planned				
China	100	100	0.2	250
USSR	10000	10000	18.1	23800
Total	**10100**	**10100**	**18.3**	**24050**
TOTAL	**53270**	**55107**		**115989**

This table includes estimates of illicit production. The estimates of industrial versus gem/near gem proportions are by Argyle Diamond Sales Ltd.

SYNTHETIC DIAMOND PRODUCTION, 1989-90, and PRODUCTIVE CAPACITY, 1990
('000 carats and % of total 1990)

	Production 1989	Production 1990	% of Production 1990	Productive Capacity 1990
Developed				
France	3100	3200	1.2	4000
Greece	800	800	0.3	1000
Ireland	58000	61000	22.6	90000
Japan	32000	34000	12.6	34000
South Africa	14100	14000	5.2	60000
Sweden	12500	12100	4.5	25000
USA	88400	89500	33.1	100000
Yugoslavia	3100	2800	1.0	5000
Total	**212000**	**217400**	**80.4**	**319000**
Developing				
Sierra Leone	-	-		10000
Centrally Planned				
China	19100	21700	8.0	60000
Czechoslovakia	3100	3000	1.1	5000
Romania	2900	2500	0.9	5000
USSR	30500	25800	9.5	70000
Total	**56000**	**53000**	**19.6**	**140000**
TOTAL	**267600**	**270400**		**459000**

Source:Argyle Diamond Sales Ltd(production estimates); U.S. Bureau of Mines (capacity estimates).

RESERVE/PRODUCTION RATIOS

Static Reserve Life (years):	18
Ratio of identified reserve base to cumulative demand 1991-2010:	0.7, but this excludes synthetic diamond and other resources

INDUSTRIAL DIAMONDS

CONSUMPTION

	Natural & Synthetic million carats		% p.a. growth rates	
	1989	1990	1970s	1980s
W Europe	89	91	n.a	n.a
Japan	40	42	13.9	5.9
USA	88	89	7.4	7.2
Other Western World	31	32	n.a	n.a
Total Western World	**248**	**254**	**n.a**	**n.a**
Eastern Countries	59	58	n.a	n.a
Total World	**306**	**312**	**n.a**	**n.a.**

Source:Argyle Diamond Sales

END USE PATTERNS, 1990 (USA) (%)

Machinery	27
Abrasives	16
Transport equipment	6
Contract construction	13
Stone and ceramic products	17
Mineral services* (drilling bits, etc)	18
Other	3

*But accounts for 59% of consumption of industrial diamond stones.

VALUE OF ANNUAL PRODUCTION

$464 million (at average US import value in 1990 for stones and grit, powders etc). Covers both natural and synthetic products.

SUBSTITUTES

Most substitutes, natural, corundum, and manufactures of fused aluminium oxide, are not as efficient or as adaptable. New abrasive materials are being brought into operation and of these cubic boron nitride seems the most promising.

TECHNICAL POSSIBILITIES

Further development of intermediate and large size industrial synthetic stones, suitable for all uses now served by natural stones.

Potential for increased markets in drilling, with replacement of conventional bits with diamond bits, in construction (diamond saws and core drills), and for diamond abrasives in the stone, glass and clay industries.

Possible new uses in electronic and electrical applications; in surgical tools and equipment; and, through electroplated metal on diamond surfaces, for bearings and protective coatings.

PRICES AND MARKETING ARRANGEMENTS

	1986	1987	1988	1989	1990	1991
US Import values						
Industrial diamond stones $/carat	7.23	10.86	9.31	6.94	6.57	7.45
Stones Real Dec 1991 prices	8.38	12.26	10.16	7.23	6.58	7.42
Bort/powder/dust $/carat	1.25	1.28	1.03	0.72	0.76	0.81

Most diamond mines produce stones of gem quality and for industrial use, and industrial supply is controlled to a large extent by gem demand. De Beers' Central Selling Organisation (CSO) controls the bulk of the world's sales of diamonds of all types. Australia markets part of its production independently. Prices vary according to size and grade; the table gives merely a crude indication. The CSO's prices for rough diamonds (gem and industrial) have not moved in the same manner as US import values.

REAL PRICES 1979 to 1991
Industrial Diamonds, US Import Values

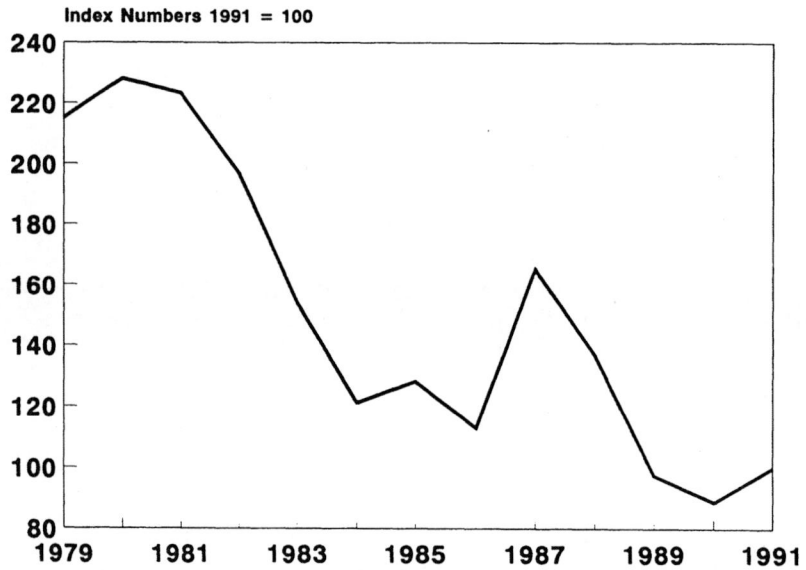

Index Numbers 1991 = 100

WORLD PRODUCTION 1979 to 1991
Industrial Diamonds, including Synthetic

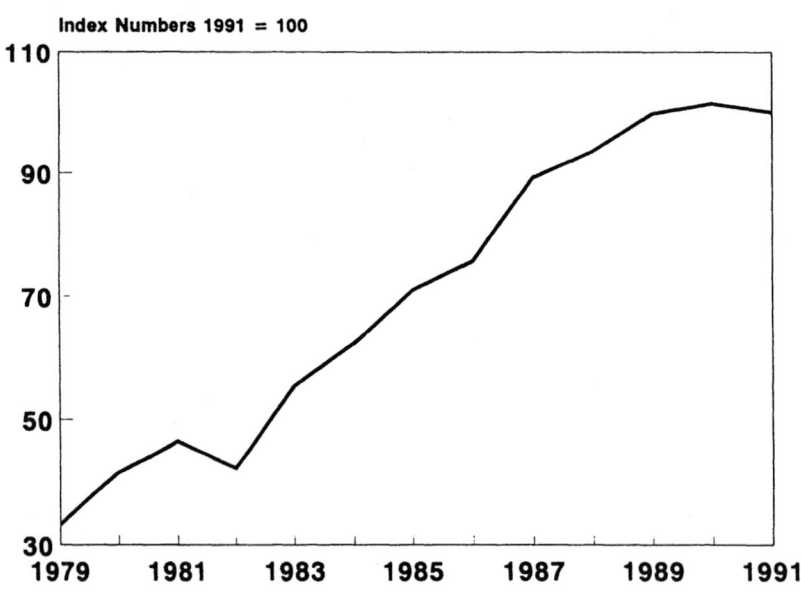

Index Numbers 1991 = 100

SUPPLY AND DEMAND BY MAIN MARKET AREA

	UK	EC(12) 1989	EC(12) 1990	Japan 1989	Japan 1990	USA 1989	USA 1990
Production ('000 carats)							
Natural	-	-	-	-	-	-	-
Synthetic	n.a	61900	65000	32000	34000	88400	89500
Secondary					3000	6000	
Net Imports ('000 carats)							
Natural	Full figures are not available			2727	2575	24748	24679
Synthetic				53842	59447	49373	73150
				(inc. dust & powder)			

Source of Net Imports (%)

Natural

	UK	EC(12) 1989	EC(12) 1990	Japan 1989	Japan 1990	USA 1989	USA 1990
Canada							
European Community				54	51	61	56
Japan						3	
S Africa							
Switzerland	Details are not			3	3	4	2
United States	available			19	24		
Ghana				1	2	1	4
Zaire				22	18	18	26
USSR				1	6	7	
Brazil					3	3	
Others				1	1	4	2

Synthetic

	UK	EC(12) 1989	EC(12) 1990	Japan 1989	Japan 1990	USA 1989	USA 1990
European Community				54	57	70	79
Romania						4	3
Japan	Details are not					10	4
Switzerland	available					4	
United States				45	42		
USSR				1		8	9
China						1	1
Others					1	3	2

Net Exports

	UK	EC(12) 1989	EC(12) 1990	Japan 1989	Japan 1990	USA 1989	USA 1990
('000 carats)							
Natural and Synthetic	-	n.a	n.a	109	121	74689	70026

Consumption

	UK	EC(12) 1989	EC(12) 1990	Japan 1989	Japan 1990	USA 1989	USA 1990
('000 carats)							
Natural and Synthetic	n/a	c.70000	c70000	40000	42000	88000	89000

INDUSTRIAL DIAMONDS

	UK	EC(12)		Japan		USA	
		1989	1990	1989	1990	1989	1990
Import Dependence(%)							
Imports as % of consumption	n.a	n.a	n.a.	n.a.	n.a.	100 (stones)	100
Imports as % of consumption and net exports	n.a	n.a.	n.a.	n.a.	n.a.	100 (stones)	100
						n.a. (others)	n.a.
Share of World Consumption (%)							
Natural & Synthetic All forms							
Western World (approx.)	n.a	c.28	c.28	16	17	35	35
Total World		c.23	c.22	13	14	29	29
Consumption Growth (% p.a.)							
1970s	n.a			13.9 (all types)		7.4 (all types)	
1980s	n.a	n.a.		5.9 (all types)		7.2 (all types)	

IRON ORE

WORLD RESERVES
('000 million tonnes of contained iron and % of total)

Developed			Developing			Centrally Planned		
Australia	10.2	(15.8)	Brazil	6.5	(0.1)	China	3.5	(5.4)
Canada	4.6	(7.1)	India	3.3	(5.1)	USSR	23.5	(36.5)
France	0.9	(1.4)	Liberia	0.5	(0.8)	Others	0.3	(0.5)
S Africa	2.5	(3.9)	Venezuela	1.2	(1.9)			
Sweden	1.6	(2.5)	Others	1.3	(2.0)			
USA	3.8	(5.9)						
Others	0.7	(1.1)						
Totals	24.3	(37.7)		12.8	(19.9)		27.3	(42.4)
Grand Total		66.1						

World reserves amount to some 151 billion tonnes of crude ore. The estimated reserve base is 101 billion tonnes of contained iron (229 billion tonnes of crude ore) and resources exceed 800 billion tonnes of crude ore with an iron content of over 230 billion tonnes

WORLD MINE PRODUCTION, 1989-90
(million tonnes of contained iron and % of total 1990)

Developed	1989	1990	% 1990	Developing	1989	1990	% 1990	Centrally Planned	1989	1990	% 1990
Australia	67.7	70.5	(13.0)	Algeria	1.3	1.3	(0.2)	China	50.0	54.0	(10.0)
Canada	23.9	22.1	(4.1)	Brazil	102.3	102.6	(18.9)	N Korea	4.4	4.4	(0.8)
France	2.9	2.6	(0.5)	Chile	5.8	5.4	(1.0)	USSR	134.8	132.0	(24.3)
New Zealand	1.1	1.2	(0.2)	Iran	1.3	1.3	(0.2)	Others	1.9	1.7	(0.3)
Norway	1.5	1.3	(0.2)	India	32.1	33.6	(6.2)				
S Africa	18.8	19.1	(3.5)	Egypt	1.5	1.5	(0.3)				
Spain	2.1	1.3	(0.2)	Liberia	7.4	2.5	(0.5)				
Sweden	14.1	12.9	(2.4)	Mauritania	7.9	7.3	(1.3)				
Turkey	1.9	2.8	(0.5)	Mexico	5.3	5.3	(1.0)				
USA	37.4	35.7	(6.6)	Peru	3.0	2.1	(0.4)				
Yugoslavia	1.7	1.4	(0.3)	Venezuela	12.0	12.7	(2.3)				
Others	2.9	2.6	(0.5)	Others	1.5	1.5	(0.3)				
Totals	176.0	173.5	(32.0)		181.4	177.1	(32.6)		191.1	192.1	(35.4)
Grand Totals	1989-	548.5									
	1990-	542.7									

IRON ORE

The gross production of ore from which the above totals were derived was 926 million tonnes in 1989 and 919 million tonnes in 1990. The average grade of ore mined was thus 59% in 1989 and in 1990. Average % grades were as follows in 1990 in the leading producing countries:

Australia	63	S Africa	63
Brazil	66	Sweden	65
Canada	61	USA	63
China	50	USSR	56
India	63	Venezuela	62
Liberia	61		
Mauritania	63		

PRODUCTIVE CAPACITY, 1990
(million tonnes of contained iron in finished iron ore products)

Developed		Developing		Centrally Planned	
Australia	77	Brazil	120	China	72
Canada	28	India	38	USSR	188
S Africa	19	Liberia	2	Others	7
Sweden	21	Mauritania	6		
USA	44	Mexico	8		
Others	15	Venezuela	20		
		Others	27		
Totals	**204**		**221**		**267**
Grand Total	**692**				

Source:Based on World Capacity & Production Report. Iron Ore Products 1991. James F King.

RESERVE/PRODUCTION RATIOS

Static Reserve Life (years):	119
Ratio of identified reserve base to cumulative demand 1991-2010:	7.9 : 1

CONSUMPTION

	million tonnes Fe content		% p.a. growth rates	
	1989	1990	1970s	1980s
European Community	92.0	83.6	-0.1	-0.2
Japan	80.5	78.9	7	-1.2
United States	50.9	48.7	-1.9	-3.7

END USE PATTERN 1990 (USA) (%)

Blast furnaces	98
Sintering plants	1
Steel furnaces, DRI + other steel functions	1
Cement production, heavy media materials and others	1

VALUE OF CONTAINED METAL IN ANNUAL PRODUCTION

$18 billion (at 1991 average prices).

SUBSTITUTES

No substitutes for steelmaking although increasing quantities of scrap are used. The main substitution comes from the replacement of steel.

TECHNICAL POSSIBILITIES

An increasing use of direct reduction processes is expected to lead to higher steel production in developing countries. Development of coal based processes for induration and direct smelting of ore would also increase efficiency.

MARKETING ARRANGEMENTS

Brazil accounts for 28%, and Australia for 25% of the world's total exports, and nine countries, each with exports of 10 million gross tonnes or more, provide 92%. Some 60% of total world production is from government controlled companies.

Captive relationships, where steel companies own and operate iron ore mines, are important in US, Canada and Australia especially. Low grade producers in N America and Europe have found competition increasingly difficult with higher grade producers in Australia and Brazil. The latter countries dominate the market but have divergent interests and objectives which limit the prospects of any agreement between them to regulate the market.

IRON ORE

PRICES

	1986	1987	1988	1989	1990	1991
US cents/Fe unit						
Europe(DMT basis)						
Brazil						
-CVRD fines (fob)	26.26	24.5	23.5	26.56	30.80	33.25
-CVRD pellets (fob)	35.6	36.7	40.35	47.33	51.60	52.15
Australia						
-Hamersley fines (cif)	32.4	29.35	31.35	35.3	41.47	41.90
-Hamersley lump (cif)	36.2	33.15	36.0	43.0	49.97	50.25
Japan(fiscal years DLT basis)						
Brazil						
-CVRD fines (fob)	23.66	22.24	21.23	23.99	27.82	30.03
Australia						
-Hamersley fines (fob)	25.97	24.67	23.68	26.76	31.03	33.49
-Hamersley lump (fob)	30.29	28.78	28.78	33.76	39.15	41.48
$/tonne						
Brazil 65% Fe cif North Sea Ports	21.89	22.23	23.12	26.50	30.80	33.25
Real Dec 1991 prices	25.41	27.48	25.22	27.59	30.83	33.12

Most prices are fixed annually under long term sales contracts although the spot market temporarily became more important during the recession-hit 1980s. Prices are influenced by the supply/demand conditions in the steel industry prevailing at time of renegotiation and they tend to lag behind economic activity. Freight is a major component of price. The North American market is largely insulated from international developments by freight rates. There are wide price ranges depending on grade and nature of product. Two reference prices tend to dominate the international market: the delivered prices of Brazilian ore to W Europe and of Australian ore to Japan. These two prices have become increasingly closely linked in recent years. Developments in one area are quickly transmitted to the other.

WORLD PRODUCTION 1979 to 1991
Iron Ore

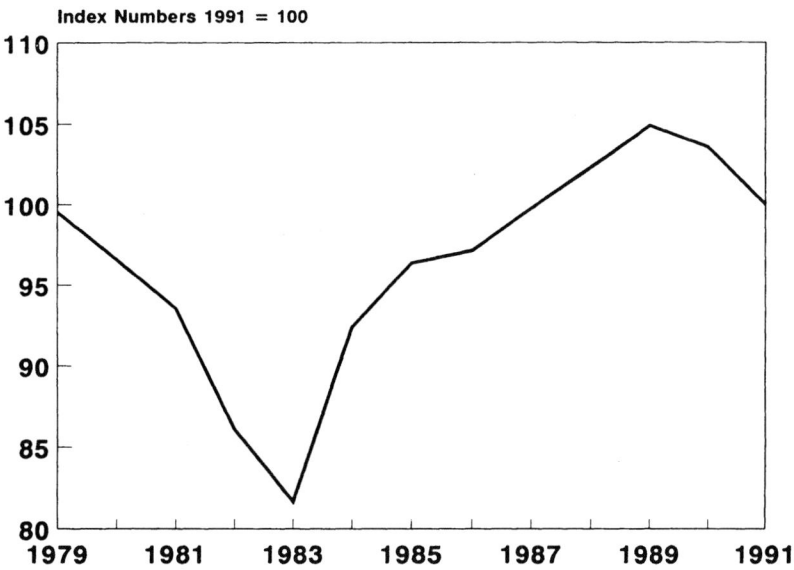

REAL PRICES 1979 to 1991
Iron Ore, Brazilian in N.W.Europe

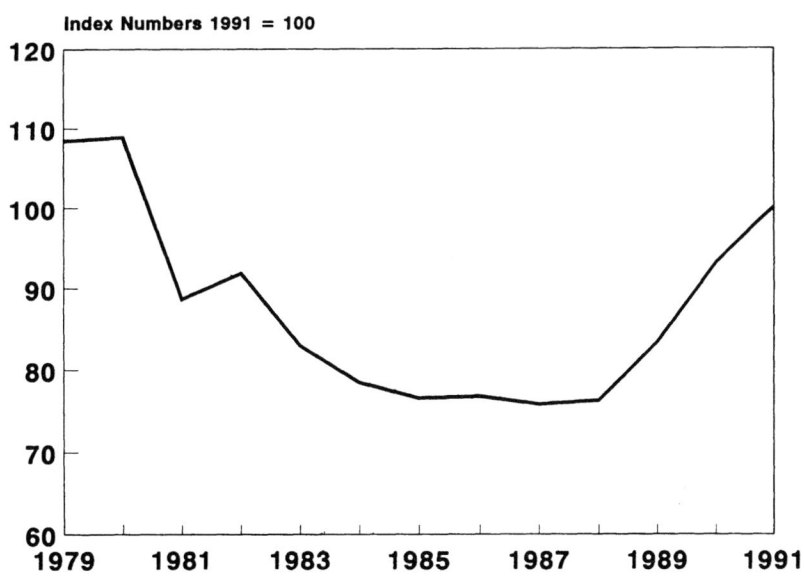

IRON ORE

SUPPLY AND DEMAND BY MAIN MARKET AREA

	UK		EC(12)		Japan		USA	
	1989	1990	1989	1990	1989	1990	1989	1990
Production								
(million tonnes)	0.03	0.05	16.0	13.88	..	..	59.0	56.4
Fe content	..	0.01	5.91	4.86	..	..	37.4	35.7
Net Imports (million tonnes)								
Iron Ores & Concentrates	19.18	14.70	135.9	124.3	127.7	125.3	19.48	18.08
Fe content approx.	11.9	9.1	85.6	78.3	80.5	78.9	12.3	11.4
Roasted iron pyrites	0.28	0.24	0.50	0.39	n.a	n.a	0.12	-
Source of Total Net Imports (%)								
Australia	25	23	16	15	44	43	2	
Canada	27	18	13	12	2	2	44	52
European Community	3	4						
Norway	5	10	3	5				
S Africa	12		5	4	4	4		
Sweden	3	3	7	7	1			
New Zealand				1				
Brazil	16	16	33	37	23	24	26	24
Chile			1	1	4	4		
India			2	2	17	17		
Liberia			8	4				
Mauritania	4	5	7	7			3	4
Philippines				4	4			
Venezuela	6	7	5	5		1	22	19
Others	2	14		1		2	3	1
Net Exports (million tonnes)								
Iron ores & concentrates	-	-	0.05	0.41	-	-	5.35	3.18
Roasted iron pyrites	-	-	0.01	0.03	n.a	n.a	-	-
Consumption								
(million tonnes)	18.60	17.99	152	137.8	127.7	125.3	80.4	76.9
Fe content approx.	11.5	11.1	92.0	83.6	80.5	78.9	50.9	48.7
Import Dependence (Fe content)								
Imports as % of consumption	100	82	93	94	100	100	24	23
Imports as % of consumption and net exports	100	82	93	94	100	100	23	22
Share of World Consumption (%)								
Total world (approx.)	2	2	17	15	15	15	9	9
Consumption Growth (% p.a.)								
1970s	-2.2		-0.1		7.0		-1.9	
1980s	1.3		-0.2		-1.2		-3.7	
(NB: 1980 strike)								

KAOLIN

WORLD RESERVES
(million and % of total)

Developed			Developing			Centrally Planned		
Australia	455	(2.3)	Brazil	1300	(6.6)	Bulgaria	700	(3.6)
Canada	150	(0.8)	India	1000	(5.1)	China	200	(1.0)
S.Africa	225	(1.9)	Iraq	210	(1.1)	USSR	200	(1.0)
Spain	150	(0.8)	Tanzania	100	(0.5)			
UK	1815	(9.2)	Others	4430	(22.5)			
USA	7175	(36.4)						
Others	1580	(8.0)						
Totals	11550	(58.6)		7040	(35.8)		1100	(5.6)
Grand Total		19690						

The table, which covers identified resources, is derived from The Economics of Kaolin 1990. Roskill Information Services.

WORLD PRODUCTION, 1989-90
('000 tonnes and % of total 1990)

Developed	1989	1990	% 1990	Developing	1989	1990	% 1990	Centrally Planned	1989	1990	% 1990
Australia	185.0	200.0	(0.8)	Argentina	185.0	150.0	(0.6)	Bulgaria	220.0	186.5	(0.7)
Austria	157.3	155.0	(0.7)	Brazil	990.0	975.0	(3.9)	Czecho-			
France	1400.0	1400.0	(5.6)	Chile	58.0	32.4	(0.1)	slovakia	698.0	670.5	(2.7)
W.Germany	777.0	780.0	(3.1)	Columbia	540.0	540.0	(2.2)	E.Germany	150.0	115.0	(0.5)
Japan	157.7	164.8	(0.7)	Malaysia	108.3	153.0	(0.6)	Hungary	24.8	18.0	..
S.Africa	139.7	132.4	(0.5)	Mexico	141.5	132.0	(0.5)	N.Korea	1239.2	146.6	(5.8)
Spain	440.0	435.0	(1.7)	Taiwan	98.1	106.0	(0.4)	Romania	400.0	400.0	(1.6)
Turkey	257.4	230.0	(0.9)	Thailand	328.8	347.7	(1.4)	USSR	2000.0	1800.0	(7.2)
UK	3140.0	3037.0	(12.1)	Venezuela	15.0	12.0	(..)	Others	50.0	50.0	(0.2)
USA	8973.7	9761.8	(38.9)								
Yugoslavia	200.0	210.0	(0.8)								
Others	372.6	350.5	(1.4)								
Totals	16118.4	6856.5	(67.2)	3514.1	3536.9	(14.1)		4782.0	4686.6	(18.7)
Grand Totals	1989-	2451.4									
	1990-	25082									

KAOLIN

PRODUCTIVE CAPACITY, 1990
('000 tonnes)

Developed		Developing		Centrally Planned	
Australia	230	Argentina	140	Bulgaria	300
Austria	110	Brazil	820	Czecho-	
				slovakia	730
France	1450	Chile	54	E.Germany	650
W.Germany	180	Columbia	1360	Hungary	41
Japan	230	Malaysia	115	N.Korea	910
S.Africa	200	Mexico	270	Romania	450
Spain	500	Taiwan	90	USSR	3100
Turkey	250	Thailand	275	Others	55
UK	3700	Venezuela	27		
USA	8955	Others	1383		
Yugoslavia	270				
Others	448				
Totals	**16523**		**4534**		**6236**
Grand Total	**27293**				

RESERVE/PRODUCTION RATIOS

Static Reserve Life (years):	Very large
Ratio of identified reserve base to cumulative demand 1991-2010:	33: 1(the reserve base ratio is much lower)

CONSUMPTION

	'000 tonnes		% p.a. growth rates	
	1989	1990	1970s	1980s
European Community	c.5100	c.5000	n.a	n.a
Japan	c.1200	c.1300	n.a	n.a
United States	7360	c.6950	n.a	4.7

END USE PATTERNS, 1990(%)

	USA (1990)	World (1988)
Paper coating	35)	43
Paper filling	21)	
Refractories	11	11
Ceramics & glass	10	5
Heavy clay products	7	15
Paint	3	6
Rubber	4	5
Others	9	15

VALUE OF ANNUAL PRODUCTION

$2.5 billion (at average 1991 prices)

SUBSTITUTES

Other clays and filler/extender minerals compete with kaolin and can be substituted at some loss of performance. In paper filling, talc and calcium carbonate may be used.

TECHNICAL POSSIBILITIES

Demand will be affected by changes within the paper industry, including the move away from acid papers and the establishment of precipitated calcium carbonate plants at US paper mills. Increased use of light weight coated papers and computer papers.

PRICES

	1986	1987	1988	1989	1990	1991
Coating clay ex UK mine, refined bagged /tonne	07.5	97.5	97.5	97.5	97.5	97.5
Filler clay ex UK mine, refined bagged /tonne	50	50	50	50	52.5	55
Filler clay $/tonne Real Dec 1991prices	85.1	92.6	97.2	85.3	94.2	96.9

MARKETING ARRANGEMENTS

Production and trade are dominated by English China Clays in the UK and USA and by Engelhard Corporation in the USA. There are additionally regional or local producers.

Prices vary widely with technical specifications and end uses.

REAL PRICES 1979 to 1991
Kaolin, Filler Clay ex UK Mine

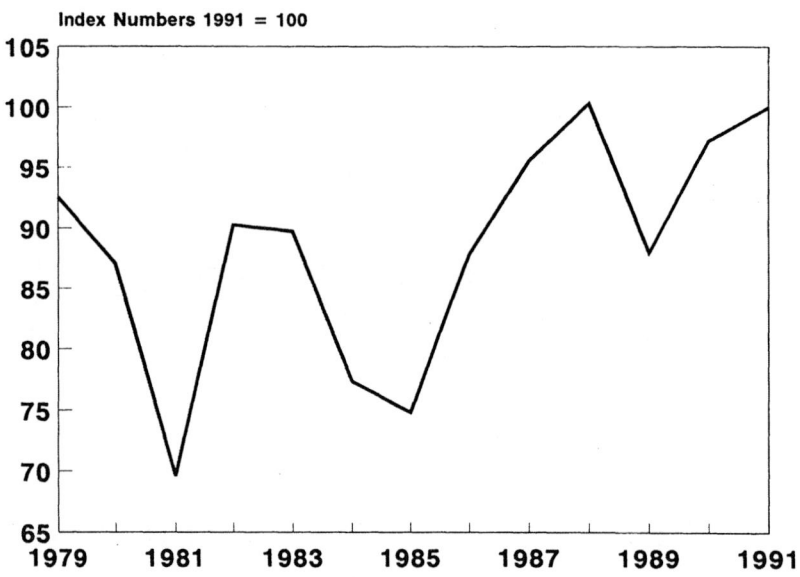

Index Numbers 1991 = 100

WORLD PRODUCTION 1979 to 1991
Kaolin

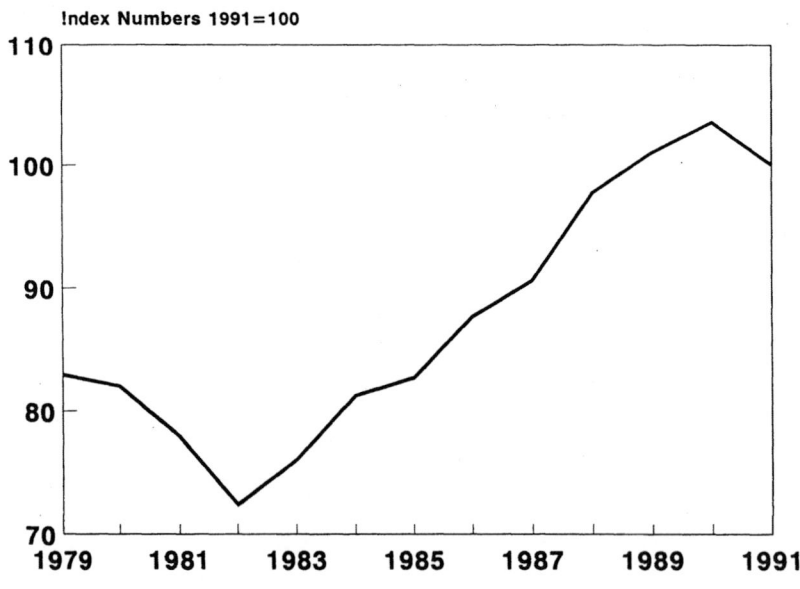

Index Numbers 1991=100

SUPPLY AND DEMAND BY MAIN MARKET AREA

	UK 1989	UK 1990	EC(12) 1989	EC(12) 1990	Japan 1989	Japan 1990	USA 1989	USA 1990
Production ('000 tonnes)	3140	3037	5757	5652	158	165	8974	9762
Net Imports ('000 tonnes)	9.1	16.9	943	917	1030	1168	3.5	3.5
Source of Net Imports (%)								
European Community	69	56			1	1	91	88
USA	28	40	55	54	64	67		
Brazil			12	13	10	9		
China					4	4		8
S. Korea					4	3		
Malaysia					3	2		
Indonesia					4	3		
Australia					8	10		
New Zealand					1	1		1
Czechoslovakia			22	22				
W. Germany			2	1				
Senegal			4	5				
Austria			1	1				
Others	3	4	4	4	1		9	3
Net Exports ('000 tonnes)	2708.1	2596.3	1584	1560	7.1	3.2	2337	2826
Consumption ('000 tonnes)	c.440	c.460	c.5100	c.5000	c.1200	c.1300	7360	c.6950
Import Dependence Imports as % of consumption	-	-	18	18	86	90	-	-
Imports as % of consumption and net exports	-	-	14	14	86	90	-	-
Share of World Consumption (%) Total World	2	2	21	20	5	5	30	28
Consumption Growth (% p.a.)								
1970s	n.a		n.a		n.a		2.1	
1980s	n.a		n.a		n.a		4.7	

129

LEAD

WORLD RESERVES
(million tonnes of metal and % of total)

Developed			Developing			Centrally Planned		
Australia	14	(19.7)	India	2	(2.8)	Bulgaria	3	(4.2)
Canada	7	(9.9)	Mexico	3	(4.2)	China	6	(8.5)
S Africa	2	(2.8)	Morocco	2	(2.8)	Poland	2	(2.8)
Spain	2	(2.8)	Peru	2	(2.8)	USSR	9	(12.7)
Sweden	2	(2.8)	Others	1	(1.5)			
USA	11	(15.5)						
Yugoslavia	2	(2.8)						
Others	1	(1.4)						
Totals	41	(57.7)		10	(14.1)		20	(28.2)
Grand Total		71						

The reserve base is estimated at 120 million tonnes. The figures for some of the smaller countries are heavily rounded with the result that the totals for 'others' in both the developed and developing groups, which are obtained by difference, appear too low.

Total world resources are estimated at 1.4 billion tonnes.

WORLD MINE PRODUCTION, 1989-90
('000 tonnes of contained metal and % of total 1990)

Developed	1989	1990	% 1990	Developing	1989	1990	% 1990	Centrally Planned	1989	1990	% 1990
Australia	495	561	(16.9)	Argentina	27	23	(0.7)	Bulgaria	49	45	(1.4)
Canada	276	232	(7.0)	Bolivia	16	20	(0.6)	China	341	315	(9.5)
W Germany	9	9	(0.3)	Brazil	15	14	(0.4)	Czecho-			
Greece	24	26	(0.8)	Honduras	6	4	(0.1)	slovakia	3	3	(0.1)
Greenland	24	17	(0.5)	India	25	25	(0.7)	N Korea	70	60	(1.8)
Ireland	32	35	(1.1)	Iran	10	9	(0.3)	Poland	51	46	(1.4)
Italy	15	16	(0.5)	Mexico	180	180	(5.4)	Romania	24	15	(0.4)
Japan	19	19	(0.6)	Morocco	65	67	(2.0)	USSR	500	490	(14.8)
S Africa	78	70	(2.1)	Namibia	24	21	(0.6)				
Spain	63	62	(1.9)	Peru	193	198	(5.7)				
Sweden	83	84	(2.5)	S Korea	17	19	(0.6)				
USA	419	495	(14.9)	Thailand	24	22	(0.7)				
Yugoslavia	79	83	(2.5)	Zambia	12	12	(0.4)				
Others	25	21	(0.6)	Others	11	7	(0.2)				
Totals	1641	1730	(52.2)		625	612	(18.4)		1038	974	(29.4)
Grand Totals	1989-	3304									
	1990-	3316									

WORLD SMELTER PRODUCTION FROM ORES AND BULLION, 1989-90
('000 tonnes and % of total 1990)

Developed	1989	1990	% 1990	Developing	1989	1990	% 1990	Centrally Planned	1989	1990	% 1990
Australia	189	207	(6.8)	Argentina	12	5	(0.2)	Bulgaria	70	48	(1.6)
Austria	10	6	(0.2)	Bazil	32	30	(1.0)	China	269	266	(8.7)
Belgium	70	70	(2.3)	Myaaner	2	2	(0.1)	N Korea	90	80	(2.6)
Canada	157	100	(3.3)	Kenya	1	1	(-)	Poland	27	15	(0.5)
France	149	137	(4.5)	India	21	25	(0.8)	Romania	25	20	(0.7)
W Germany	170	162	(5.3)	S Korea	49	34	(1.1)	USSR	520	500	(16.4)
Greece	6	-	(-)	Mexico	174	173	(5.7)				
Italy	76	69	(2.3)	Morocco	64	65	(2.1)				
Japan	208	205	(6.7)	Namibia	44	35	(1.1)				
Spain	61	57	(1.9)	Peru	74	69	(2.3)				
Sweden	35	35	(1.2)	Zambia	4	4	(0.1)				
UK	157	156	(5.1)	Thailand	7	5	(0.2)				
USA	397	404	(13.3)								
Yugoslavia	71	58	(1.9)								
Totals	1756	1666	(54.8)		484	448	(14.7)		1001	929	(30.5)
Grand Totals	1989-	3241									
	1990-	3043									

WORLD REFINED LEAD PRODUCTION, 1989-90
('000 tonnes and % of total 1990)
This includes secondary antimonial lead.

Developed	1989	1990	% 1990	Developing	1989	1990	% 1990	Centrally Planned	1989	1990	% 1990
Australia	204	224	(3.9)	Argentina	25	20	(0.4)	Bulgaria	102	66	(1.2)
Belgium	93	92	(1.6)	Brazil	86	76	(1.3)	China	302	287	(5.1)
Canada	243	192	(3.4)	India	37	42	(0.8)	Czecho-			
France	267	260	(4.5)	Mexico	174	173	(3.0)	slovakia	26	24	(0.4)
W Germany	350	349	(6.2)	Morocco	66	67	(1.2)	E Germany	40	45	(0.8)
Italy	181	171	(3.0)	Namibia	44	35	(0.6)	N Korea	70	65	(1.1)
Japan	333	329	(5.8)	Peru	74	69	(1.2)	Poland	78	65	(1.1)
Spain	114	124	(2.2)	S Korea	106	80	(1.4)	Romania	28	22	(0.4)
Sweden	75	76	(1.3)	Taiwan	58	27	(0.5)	USSR	750	730	(12.9)
UK	350	329	(5.8)	Others	112	113	(2.0)				
USA	1253	1292	(22.8)								
Yugoslavia	119	94	(1.7)								
Others	143	136	(2.4)								
Totals	3725	3668	(64.6)		782	702	(12.4)		1396	1304	(23.0)
Grand Totals	1989-	5903									
	1990-	5674									

LEAD

MINE, SMELTER AND REFINERY CAPACITIES, 1990
('000 tonnes)

	Mine	Smelter	Refinery
Developed			
Australia	590	468	258
Austria	7	49	49
Belgium	-	115	155
Canada	381	309	308
France	1	272	272
W Germany	7	343	448
Italy	18	255	235
Japan	28	395	431
Netherlands	-	55	55
S Africa	102	39	39
Spain	72	109	122
Sweden	95	114	70
Turkey		42	42
UK	-	199	389
USA	759	1092	1180
Yugoslavia	112	131	131
Others	90	16	43
Total	**2272**	**4003**	**4100**
Developing			
Argentina	35	119	112
Bolivia	18	27	27
Brazil	28	109	109
India	44	75	75
Mexico	210	303	323
Morocco	74	68	68
Namibia	40	75	75
Peru	212	94	91
S Korea	13	-	35
Zambia	16	33	18
Others	68	46	25
Total	**758**	**949**	**952**
Centrally Planned (a)			
China	370	250	250
Bulgaria	100	130	120
N Korea	85	90	70
Poland	50	90	90
USSR	550	635	675
Total	**1155**	**1195**	**1205**
TOTAL	**4185**	**6147**	**6257**

(a) Excluding secondary smelter and refinery capacity. Smelter and Refinery Capacities include secondary metal. Primary smelting capacity was about 4.5 million tonnes in 1990 and primary refining capacity 4.56 million tonnes.

LEAD RECOVERED FROM SCRAP: WESTERN WORLD
('000 tonnes 1989-90)

	1989	1990
Scrap included in refined production	2267	2256
Other identified scrap recovery (remelted, alloys and direct use)	170	168
Total	**2437**	**2424**

RESERVE/PRODUCTION RATIOS

Static Reserve Life (years):	20
Ratio of identified reserve base to cumulative demand 1991-2010:	1.1 : 1

CONSUMPTION OF REFINED METAL

	'000 tonnes		% p.a. growth rates		
	1989	1990	1960s	1970s	1980s
European Community	1478	1513	2.4	0.2	0.6
Japan	406	417	8.4	3.4	0.6
United States	1346	1312	2.1	0.6	1.8
Others	1297	1178	5.8	2.7	1.9
Total Western World	**4527**	**4420**	**3.4**	**1.2**	**1.3**
Total World	**5844**	**5609**	**3.9**	**1.8**	**0.5**

END USE PATTERNS, 1990 (%)

	USA[1]	UK[1]	W Germany[1]	Japan[2]
Batteries	80	31	52	70
Cable sheathing	2	3	3	1
Pipe and sheet	2	30	13	3
Chemicals	4	22*	27	15
Alloys	4	6	2	5
Others	8*	8	2	6

* Including tetraethyl
1 Refined lead and direct use of scrap
2 Refined lead only

Source: ILZSG

LEAD

VALUE OF CONTAINED METAL IN ANNUAL PRODUCTION

$3.2 billion (total refined metal at average 1991 prices).

SUBSTITUTES

Battery replacements include nickel-zinc, zinc-chloride and lithium metal-sulphide and nickel-hydride although large scale commercial use is so far precluded by cost and operating problems.

Polyethylene and other materials substitute in some cable coverings.

In construction applications, plastics, galvanised steel, copper and aluminium are alternatives. In corrosive chemical environments, stainless steel, titanium, plastics and cement are substitutes. Tin, glass, plastics and aluminium are alternatives in tubes and containers, and iron or steel in shot for ammunition.

TECHNICAL POSSIBILITIES

Environmental concerns are limiting uses for lead particularly in petrol where its use as an anti-knock additive is rapidly being phased down, a process hastened by the introduction of catalytic converters. Storage batteries for industrial load levelling, mains power management, and electric vehicles are growing markets. Also the continued search for weight reduction is reducing the amount of lead per battery, and battery lives are being extended. Possible new developments include the use of lead as an antioxidant in asphalt, as a shielding material both in nuclear waste in protection of buildings against radon gases and as a sound baffler. Environmental legislation is, however, likely to inhibit the growth of new uses.

New techniques to recover lead from concentrates and from scrap are developing and will become more important in the 1990s.

PRICES

	1986	1987	1988	1989	1990	1991
c/lb						
US Producer	22.3	36.0	37.4	39.6	46.0	33.5
LME Cash	18.4	27.1	29.8	30.6	36.7	25.3
Real Dec 1991 prices	21.4	30.6	32.5	31.9	37.0	25.2
£/tonne						
LME Cash	277.1	363.4	367.9	412.3	459.2	314.2
Monthly LME	236.3-	287.5-	332.8-	336.5-	312-	274.3-
range /tonne	380.5	540.5	411.5	489.5	812.5	363.5

Outside the US, where a domestic producer pricing system operates, sales are based mainly on LME terminal market prices. A substantial percentage of mine output is associated with zinc, copper and silver. Production of these influences both the supply of lead and its breakeven costs. Large secondary production is a major factor influencing supply and prices, but it is increasingly circumscribed by environmental regulations, particularly on the transport and handling of wastes.

MARKETING ARRANGEMENTS

Some 300-400 mines produce lead mainly as by- or co-product, but smelters are the main influence on market trends. Primary smelting is dominated by large companies, with state controlled production, e.g. from Peru, a growing influence. Secondary smelters, often linked to battery manufacturers, normally have a restraining effect on the market; scrap availability is fairly sensitive to price.

REAL PRICES 1979 to 1991
Lead, LME Cash

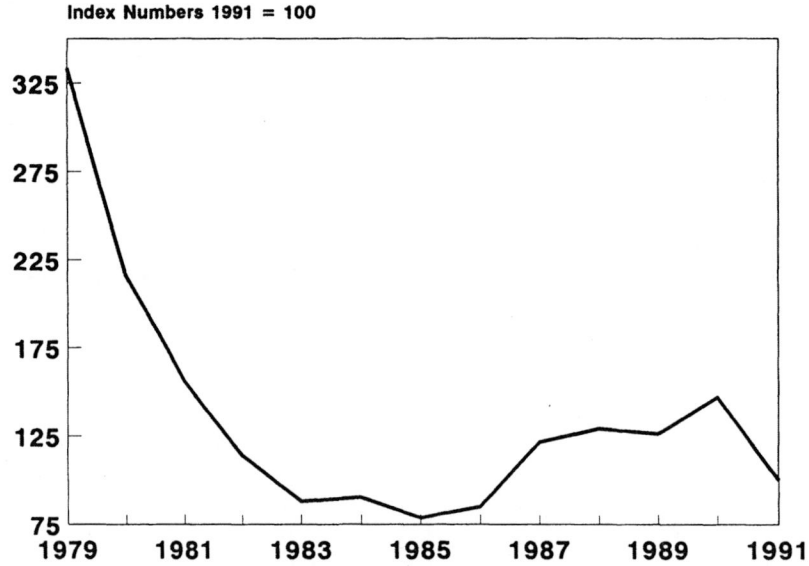

Index Numbers 1991 = 100

WORLD PRODUCTION 1979 to 1991
Lead, Refined Metal

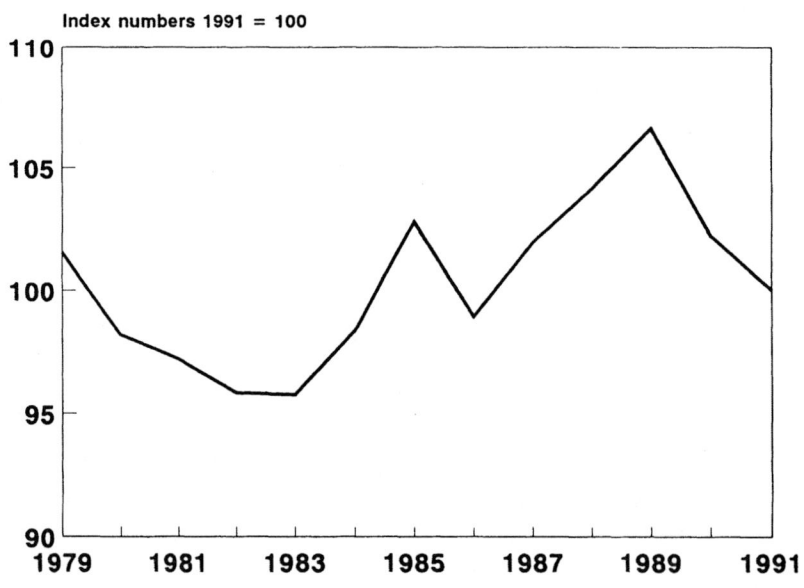

Index numbers 1991 = 100

SUPPLY AND DEMAND BY MAIN MARKET AREA

	UK		EC(12)		Japan		USA	
	1989	1990	1989	1990	1989	1990	1989	1990
Production								
('000 tonnes)								
Mine production	2.2	1.4	146.4	149.8	18.6	18.77	419.3	495.2
Smelted from ores & bullion	156	156	689	651	208	205	396	404
Refined inc. secondary	350	329	1422	1389	333	329	1253	1291
Net Imports								
('000 tonnes lead content)								
Ores and concentrates	30.8	40.7	332.1(a)	291.6(a)	178.5	178.1	2.9	7.8
Base bullion	164.3	156.7	168.7	170.0	2.2	1.6	5.2	2.9
Refined inc. refined antimonial	53.1	31.8	168.1	172.3	76.1	91.3	115.7	90.6

(a) exc. Greece

Source of Net Imports (%)

Ores and concentrates

	UK		EC(12)		Japan		USA	
	1989	1990	1989	1990	1989	1990	1989	1990
Australia	25	18	16	11	24	27	1	
Canada	3	10	21	21	37	27	93	97
European Community	29	15						
Poland			1	4				
Norway			1	1				
S Africa			14	12	6	8		
Sweden	1	5	14	12				
USA	9	24	1	5	2	2		
Argentina			1	1				
Bolivia	1	4	4	1	1			
Honduras	12	4	1	1				
Mexico		2	1					
Morocco			4	7				
Peru	13	18	12	12	16	24	6	2
Thailand		1	2	2	7	6		
Turkey			1	1		2		
China						5	1	
Others	7	3	6	6	2	2		1

137

LEAD

	UK 1989	UK 1990	EC(12) 1989	EC(12) 1990	Japan 1989	Japan 1990	USA 1989	USA 1990
Refined Lead								
Australia	18	17	7	7	15	11		9
Austria			3	1				
Canada	47	45	18	11	7	9	63	61
European Community	16	3				3		
S Africa & Namibia			7	6				
Sweden	10	18	12	15				
United States			3	1	2			
Chile					1			
S Korea				1	8			
Mexico	6		7	10	29	18	26	28
Morocco			23	17	1			
Peru			8	9	12	20	6	1
Taiwan				11	6			
Czechoslovakia			3	1				
China					10			
North Korea				6	5			
USSR		11	5	7	11	8		
Others	3	6	7	13	3	2	2	1
Indonesia				6				

Net Exports
('000 tonnes lead content)

	UK 1989	UK 1990	EC(12) 1989	EC(12) 1990	Japan 1989	Japan 1990	USA 1989	USA 1990
Ores and concentrates	0.6	1.4	34.1	12.8	-	-	57.0	56.6
Base bullion	26.6	24.8	13.5	47.6(a)	1.3	1.4	3.1	1.7
Refined lead, inc. refined antimonial	66.4	77.2	102.4	60.6	0.1	0.1	28.5	57.2

(a)Gross exports

Consumption
('000 tonnes)

	UK 1989	UK 1990	EC(12) 1989	EC(12) 1990	Japan 1989	Japan 1990	USA 1989	USA 1990
Refined	301.3	301.6	1477.7	1512.7	)405.7	416.9	1341.1	1311.7
Scrap and remelted (not included in refined)	35.0	32.5	79.5	75.2	)		34.7 (excludes remelted)	35.0

Import Dependence

	UK 1989	UK 1990	EC(12) 1989	EC(12) 1990	Japan 1989	Japan 1990	USA 1989	USA 1990
Imports as % of consumption	82	76	45	42	63	65	9	8
Imports as % of consumption and net exports	63	57	41	39	63	65	9	7

Share of World Consumption (%)
(Refined Lead)

	UK 1989	UK 1990	EC(12) 1989	EC(12) 1990	Japan 1989	Japan 1990	USA 1989	USA 1990
Western world	7	7	33	34	9	9	30	30
Total world	5	5	25	27	7	7	23	23

Consumption Growth (% p.a.)

	UK		EC(12)		Japan		USA	
1960s	-0.1		2.4		8.4		2.1	
1970s	-0.6		0.2		3.4		0.6	
1980s	0.2		0.6		0.6		1.8	

LITHIUM

WORLD RESERVES
('000 tonnes lithium and % of total)

Developed			Developing			Centrally Planned	
Australia	372	(16.8)	Brazil	1	(0.1)	China	n.a
Canada	181	(8.2)	Chile	1270	(57.5)	USSR	n.a
USA	363	(16.4)	Zimbabwe	23	(1.0)		
Totals	**916**	**(41.4)**		**1294**	**(58.6)**		**n.a**
Grand Total	**2210 (Western World)**						

The western world reserve base, in so far as data are available, is estimated at 8.35 million tonnes, two-thirds of which is in Bolivia. Another location, not included above, is Zaire. The size of the reserve base in Argentina, Brazil, Namibia and Portugal is unknown. Total estimated world resources are approximately 12.7 million tonnes of lithium equivalent.

WORLD MINE PRODUCTION, 1989-90, and PRODUCTIVE CAPACITY, 1990
(tonnes of lithium and % of total 1990)

	Mine Production 1989	Mine Production 1990	% of Production 1990	Productive Capacity 1989(a)
Developed				
Australia	1300	1300	(13.1)	1300
Canada	440	440	(4.4)	500
Portugal	-	-	(-)	18
USA	c.4500	c.4500	(45.3)	5600
Total	**6240**	**6240**	**(62.8)**	**7418**
Developing				
Argentina	2	2	(..)	10
Brazil	54	50	(0.5)	290
Chile	1700	1800	(18.1)	1800
Namibia	27	27	(0.3)	30
Zimbabwe	410	410	(4.1)	730
Total	**2193**	**2289**	**(23.1)**	**2860**
Centrally Planned				
China	300	300	(3.0)	730
USSR	1100	1100	(11.1)	1090
Total	**1400**	**1400**	**(14.1)**	**1820**
TOTAL	**9833**	**9929**		**12098**

(a) Includes mines and chemical processing plants.
These figures represent estimates of lithium extracted from mineral concentrate and brine.

LITHIUM

RESERVE PRODUCTION RATIOS

Static Reserve Life (years):	extremely large
Ratio of identified reserve base to cumulative demand 1991-2010:	40 : 1

CONSUMPTION

The available statistics are sparse, and those below merely give broad orders of magnitude of contained lithium as concentrate.

	tonnes		% p.a. growth rates	
	1989	1990	1970s	1980s
European Community	c.1250	c.1300	n.a	n.a
Japan	c.850	c.850	11.7	5.0
United States	c.2700	c.2700	5.2	-2.2

Alternative estimates of consumption (in terms of tonnes of contained lithium) of chemicals and metal only are:

	1989	1990
N. America	2175	2215
W. Europe	1450	1490
Far East	940	1025
S.America	425	385
Others	485	425
Total Western World	**5475**	**5540**

Source:Mining Annual Review 1991
Lithium minerals, used mainly in the ceramics, glass and metallurgical industries, constitute about 25% of production. Lithium chemicals, often used in the same industries, make up the balance.

END USE PATTERNS, 1988 (USA) (%)

	USA(1988)	World(1990)
Primary aluminium	28	18
Ceramics and glass	32	65
Lubricants	24	8
Others	16	9

VALUE OF ANNUAL PRODUCTION

$210 million (at average 1991 price for lithium carbonate).

SUBSTITUTES

Sodium and potassium substitute as fluxes in ceramics and glass industries. Calcium and aluminium soaps, plus detergents and gels, are alternatives for lithium stearate in lubricants.

Zinc, magnesium, cadmium, sodium and mercury compete for the lithium anode material in batteries. Magnesium has also been successful as a deoxidiser and grain refiner in copper and iron castings.

Lithium can be removed from use in aluminium potlines by increasing the percentages of other salts.

TECHNICAL POSSIBILITIES

Use in nuclear fusion electric power reactors. Development of rechargeable lithium batteries and extensive use in fuel cells. Potential for substantial use in structural metal field, particularly in lightweight alloys and in glass applications.

Substitution of lithium for fluorine as a melting flux.

PRICES

	1986	1987	1988	1989	1990	1991
US carbonate 99% min lithium carbonate ¢//lb	150	152	155	155	166	180
Real Dec 1991 price	174	171	169	161	166	179

Lithium carbonate is 18.8% contained lithium. US producers set a domestic producer price which serves as a world reference price.

MARKETING ARRANGEMENTS

Two US companies, Cyprus Foote Mineral and FMC Corp, control the majority of the Western World's production of lithium concentrate. Although a Chilean lithium carbonate plant, which opened in 1984, has reduced the dominance of domestic US production, it is run by an affiliate of Cyprus Foote. The rapidly growing market for lithium mineral concentrates is mainly shared between Lithium Australia, Tanco in Canada, Cyprus Foote from US sources, and Bikita Minerals in Zimbabwe.

REAL PRICES 1979 to 1991
Lithium, US lithium carbonate

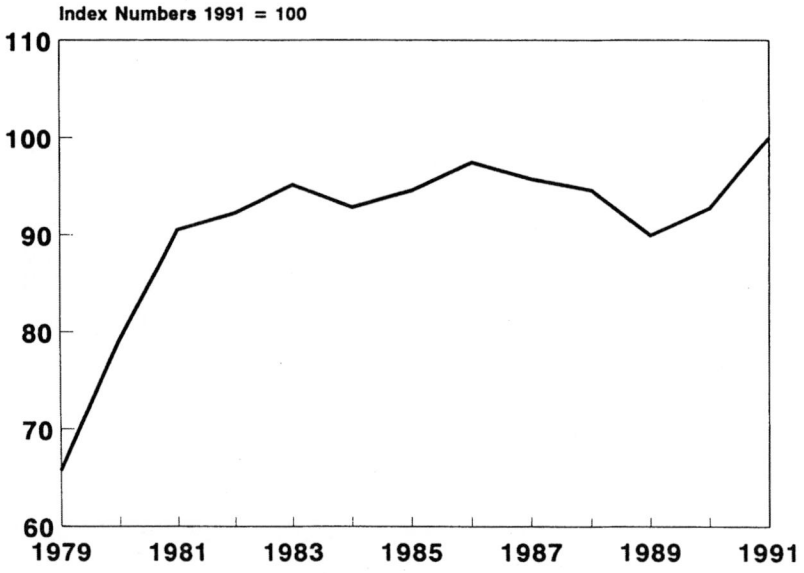

WORLD PRODUCTION 1979 to 1991
Lithium

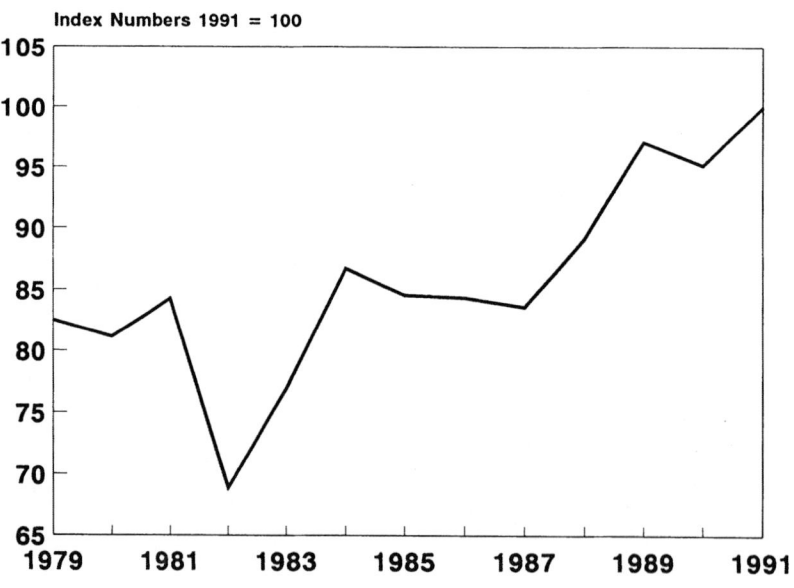

SUPPLY AND DEMAND BY MAIN MARKET AREA

	UK		EC(12)		Japan		USA	
	1989	1990	1989	1990	1989	1990	1989	1990
Production								
(tonnes)								
Mine production								
(contained Li)	-	-	-	-	-	-	c.4500	c.4500
Net Imports								
(tonnes)								
Ores and concentrates, gross	n.a	n.a	n.a	n.a	-	-		
Lithium carbonate	1976	1925	8705	8662	3933	3835	3327	4197
Lithium hydroxide	687	596	1842	1775	1020	1036	23	-
Lithium metal	n.a	n.a	n.a	n.a	n.a	n.a		
Total contained Lithium	c.485	c.460	c.1950	c.1920	c.910	c.900	630	790

(a)Excluding UK
(b)Excluding ores and concentrates

Source of Net Imports (%)

Lithium carbonate

	UK		EC(12)		Japan		USA	
European Community	15	17						
United States	83	72	67	66	43	41		
Chile			19	11	47	36	97	100
China	2	5	10	14	9	20		
USSR			3	5		2		
Others		6	1	4	1	1	3	

Lithium Hydroxide

	UK		EC(12)		Japan		USA	
European Community	21	28			2			
Switzerland							45	
United States	69	59	53	42	90	75		
China	10	12	30	47	6	8		
USSR			9	9	2	16		
Japan							41	
Others and undefined		1	8	2		1	14	

Metal

	UK		EC(12)		Japan		USA	
European Community								
United States								
Others and undefined	n.a	n.a	n.a	n.a.	n.a	n.a.		

LITHIUM

	UK		EC(12)		Japan		USA	
	1989	**1990**	**1989**	**1990**	**1989**	**1990**	**1989**	**1990**
Net Exports								
(tonnes)								
Lithium carbonate	735	591	365	442	0.8	5.4	9046	9313
Lithium hydroxide	203	164	245	195	3.4	14.5	4692	3147
Lithium metal	n.a	n.a	n.a	n.a	n.a	n.a	5.38	n.a.
Total all forms contained								
Lithium	c.175	c.140	c.110	c.115	n.a.	n.a.	3000	2600
Consumption								
(tonnes)								
Contained Lithium as								
chemicals after								
processing losses	c.310	c.320	c.1250	c.1300	c.850	c.850	2700	2700
Import Dependence								
Imports as % of consumption	100	100	100	100	100	100	23	29
Imports as % of consumption								
and net exports	100	100	100	100	100	100	11	15
Share of World Consumption (%)								
Total world	c.3	c.3	c.13	c.13	c.9	c.9	27	27
Consumption Growth (% p.a.)								
1970s	n.a		n.a		11.7		5.2	
1980s	n.a		n.a		5.0		-2.2	

MAGNESIUM

MAGNESITE - WORLD RESERVES
(million tonnes of magnesium and % of total)

Developed			Developing			Centrally Planned		
Australia	145	(5.8)	Brazil	45	(1.8)	Czecho-		
Austria	15	(0.6)	India	30	(1.2)	slovakia	20	(0.8)
Canada	30	(1.2)	Nepal	45	(1.8)	China	745	(30.0)
Greece	30	(1.2)	Others	180	(7.2)	N Korea	445	(17.9)
South Africa	5	(0.2)				Poland	10	(0.4)
Spain	10	(0.4)				USSR	650	(26.2)
Turkey	65	(2.6)						
USA	10	(0.4)						
Yugoslavia	5	(0.2)						
Totals	315	(12.7)		300	(12.1)		1870	(75.3)
Grand Total		2485						

The reserve base is 3400 million tonnes. Identified world resources of magnesite total some 12 billion tonnes. Furthermore magnesium compounds can be recovered economically from well and lake brines and from seawater. The latter, which contains 0.13% by weight of magnesium, is a major source of metal and compounds.

WORLD PRODUCTION OF MAGNESITE, 1989-90
('000 tonnes and % of total 1990)

Developed	1989	1990	% 1990	Developing	1989	1990	% 1990	Centrally Planned	1989	1990	% 1990
Austria	1205	1200	(11.2)	Brazil	260	350	(3.3)	Czecho-			
Australia	55	60	(0.6)	India	480	500	(4.7)	slovakia	600	600	(5.6)
Canada	150	150	(1.3)	Nepal	28	25	(0.2)	China	2000	2000	(18.7)
Greece	850	860	(8.0)	Zimbabwe	33	30	(0.3)	N Korea	1500	1500	(14.0)
S Africa	76	104	(1.0)	Others	37	35	(0.3)	Poland	23	24	(0.2)
Spain	620	600	(5.6)					USSR	1825	1600	(14.9)
Turkey	1238	800	(7.5)								
Yugoslavia	364	275	(2.6)								
Totals	4558	4049	(37.8)		838	940	(8.8)		5948	5724	(53.4)
Grand Totals	1989-	11344									
	1990-	10713									

The magnesium content of this production averaged approximately 3.1 million tonnes. In addition the magnesium content of dolomite, seawater, and well and lake brines averaged some 1.75 - 2 million tonnes of contained magnesium, with output in the United States 513,000 tonnes in 1989 and 499,000 tonnes in 1990(including a small magnesite based output). The world capacity for producing magnesia from seawater and brines is roughly 2.4 million tonnes (US 0.91 million, Japan 0.49 million). This is equal to 1.85 million tonnes of contained magnesium, whilst magnesite based capacity is 6.4 million tonnes of contained magnesium.

MAGNESIUM

WORLD PRODUCTION, 1989-90, and PRODUCTIVE CAPACITY, 1990, OF PRIMARY MAGNESIUM METAL
('000 tonnes and % of total 1990)

	Mine Production		% of Production	Productive Capacity
	1989	1990	1990	1990
Developed				
Canada	7.2	25.3	(7.1)	61.5
France	14.6	14.6	(4.1)	15.0
Italy	5.8	5.9	(1.6)	10.0
Japan	12.1	12.8	(3.6)	13.0
Norway	49.8	48.2	(13.5)	41.0
USA	152.1	139.3	(38.9)	179.0
Yugoslavia	6.1	5.8	(1.6)	7.0
Total	**247.7**	**251.9**	**(70.4)**	**326.5**
Developing				
India	0.3	1.0	(0.3)	0.6
Brazil	6.2	8.7	(2.4)	10.6
Total	**6.5**	**9.7**	**(2.7)**	**11.2**
Centrally Planned				
China	15.0	16.0	(4.5)	9.0
USSR	85.0	80.0	(22.4)	95.0
Total	**100**	**96.0**	**(26.9)**	**104.0**
TOTALS	**354.2**	**357.6**		**441.7**

SECONDARY RECOVERY OF MAGNESIUM METAL
('000 tonnes)

	1989	1990
Austria	0.2	0.1
Brazil	1.5	1.6
Japan	17.8	20.4
United Kingdom	0.6	1.0
United States	51.2	50.3
USSR	8.0	7.5

This includes recovery of magnesium alloys.

RESERVE/PRODUCTION RATIOS

Static Reserve Life (years):	extremely large (excludes seawater)
Ratio of identified resource base to cumulative demand 1991-2010:	over 40 : 1

(This excludes seawater, brines and presently uneconomic resources.)

CONSUMPTION OF MAGNESITE

Reliable data for most countries are not readily available. United States' consumption of magnesium compounds was 658,000 tonnes of contained magnesium in 1989 and 587,000 tonnes in 1990. It fell at average annual rates of 2% during the 1970s, and 3% in the 1980s, mainly because of reduced activity in the steel industry.

CONSUMPTION OF MAGNESIUM METAL

		'000 tonnes			% p.a. growth rates (total)	
	Primary 1989	Total 1989	Primary 1990	Total 1990	1970s	1980s
European Community		48.0		54.7	-2.0	0.1
Japan (a)	22.8	43.7	25.0	51.0	8.5	3.2
United States (a)	105.2	156.4	96.1	146.4	2.8	1.0
Other countries		56.1		61.6	3.2	3.3
Total Western World		**304.2**		**313.7**	**3.0**	**1.5**
Total World		**421.9**		**412.2**	**3.9**	**1.7**

(a)The differences between primary and total consumption represent magnesium recovered from scrap (and especially can stock, reused as such). Secondary production of magnesium metal is relatively small in the world as a whole.

END USE PATTERNS 1990, (USA)(%)

Non-metal
Refractories	75
Preparation of caustic calcined and specified magnesias and other magnesium compounds	25

Metal
Manufacture of Aluminium based alloys	47
Castings and wrought products	21
Reducing agent	9
Desulphurisation of iron & steel	10
Cathodic protection	6
Other	7

VALUE OF ANNUAL PRODUCTION

Magnesite(excl. brines)	$ 1.5 billion (at average 1991 prices)
Magnesium metal (primary only)	$ 1.13 billion (at average 1991 prices)

As magnesite is a raw material for some magnesium metal, the two values are not additive.

SUBSTITUTES

Aluminium, zinc, plastics, and composites are alternatives in many die-casting applications. Sodium can be used to reduce titanium tetrachloride to produce titanium metal.

Rare earth elements and calcium carbide can substitute in the production of nodular iron and steel to some extent in desulphurisation.
Alumina, silica, zirconia, chromite and kyanite are substitutes in magnesia refractories.

MAGNESIUM

TECHNICAL POSSIBILITIES

Increased use in the car industry and in steel desulphurisation.

Greater use of aluminium-magnesium alloys, particularly in the aluminium can, and the development of new improved alloys.

Development of better refractories and of furnaces with limited refractory maintenance could decrease demand.

Olivine and dunite, naturally occurring magnesium compounds, are potential alternatives for silica foundry sand and blasting sand. Olivine also has some limited use for slag and alkali control in steelmaking.

PRICES

	1986	1987	1988	1989	1990	1991
Magnesite, Greek crude lump cif main European port						
/tonne	55-60	55-60	55-60	55-60	-	-
$/tonne	84	94	102	94	-	-
Real Dec 1991 prices	98	106	111	98	-	-
Magnesite, Turkish. raw, max.0.6% Si. f.o.b.						
Marmara $/tonne	-	-	-	-	55-75	60-70
Real Dec. 1991 prices	-	-	-	-	65.2	62.3
Magnesium metal US primary ingot						
99.8% cents/lb	153.0	153.0	156.3	163.0	161.4	143.0
Real Dec 1991 prices	177.3	172.7	170.3	169.3	162.0	142.5

MARKETING ARRANGEMENTS

Metal production is dominated by US and Norwegian companies but sources of raw materials (seawater, lake brines, magnesite, dolomite) are widespread. Costs of energy are a limiting factor on new metal production, with present production technology.

New productive capacity has opened in Canada in 1990-91, but the US imposed a bitterly contested countervailing duty in 1992.

European Community producers of magnesite have sought protection against low priced material from China.

REAL PRICES 1979 to 1991
Magnesite, Greek and Turkish Ore

Index Numbers 1991 = 100

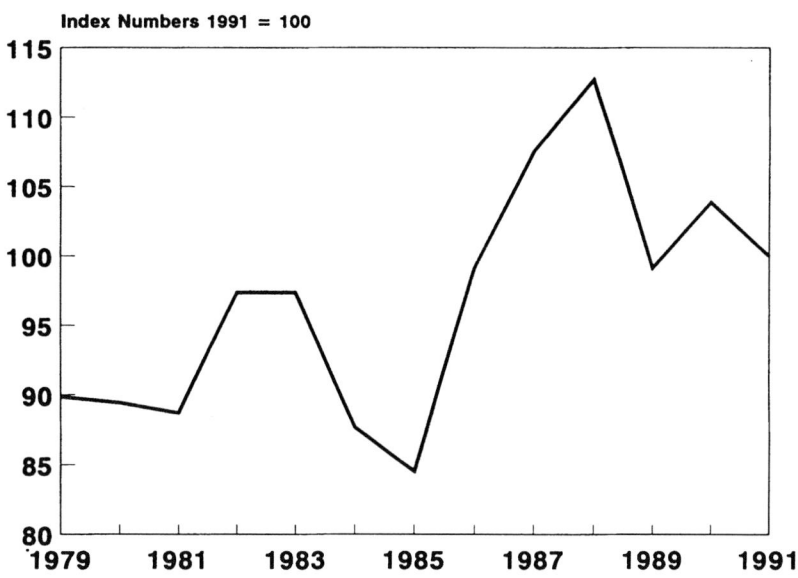

WORLD PRODUCTION 1979 to 1991
Magnesite

Index Numbers 1991 = 100

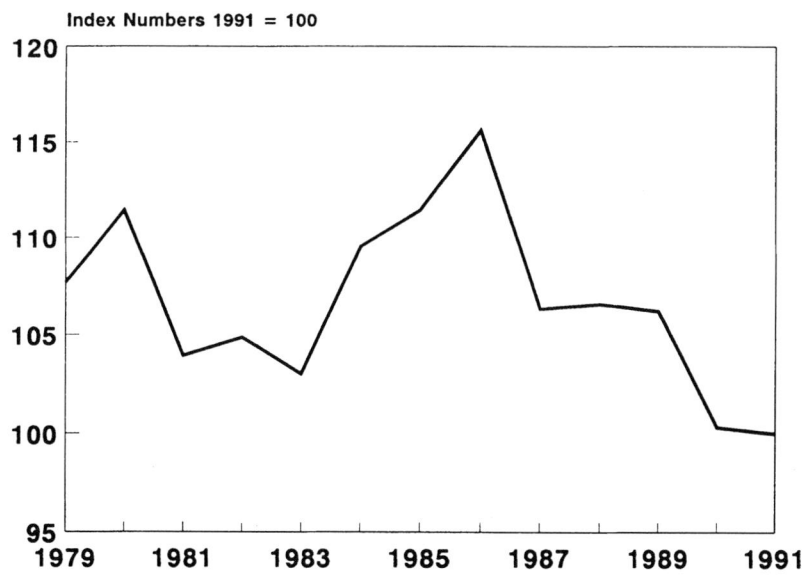

MAGNESIUM

SUPPLY AND DEMAND FOR MAGNESITE BY MAIN MARKET AREA

	UK		EC(12)		Japan		USA	
	1989	**1990**	**1989**	**1990**	**1989**	**1990**	**1989**	**1990**
Production								
('000 tonnes)								
Magnesite-gross weight	-	-	1470	1460	-	-	n.a	n.a
-Mg content	-	-	467	421	-	-	n.a	n.a
Magnesia, from other sources								
(dolomite, brines, seawater)								
-Mg content	n.a	n.a	n.a	n.a	n.a	n.a	513	499(a)
		(capacity 200)						

(a) Including small tonnages of magnesite.

	UK		EC(12)		Japan		USA	
Net Imports								
('000 tonnes)								
Magnesium oxide, carbonate								
and clinker	157	153	673	746	430	373	314	284

Source of Net Imports (%)
including magnesia from brine and seawater

Crude + processed magnesite

	UK		EC(12)		Japan		USA	
	1989	**1990**	**1989**	**1990**	**1989**	**1990**	**1989**	**1990**
Austria	2	1	14	11	..	..		
Canada				1			28	32
European Community	63	67					27	19
Japan	2		1	1			5	4
United States	1	2	2	1				
Turkey	5	4	21	17			1	
China	9	17	31	42	92	90	24	31
Czechoslovakia	5	3	9	7			4	1
N Korea	9		11	9	4	4		
Israel	2	1	4	3	2	2		3
Mexico		3	2				9	10
Yugoslavia	2	1						
Others	2	5	2	5	2	4	2	

	UK		EC(12)		Japan		USA	
Net Exports								
('000 tonnes)	90	75	201(a)	160(a)	122	95	93	117

(a) Excluding Netherlands

	UK		EC(12)		Japan		USA	
Consumption								
('000 tonnes)	n.a	n.a	n.a	n.a	n.a	n.a	658	587
							(Mg content)	

	UK		EC(12)		Japan		USA	
	1989	**1990**	**1989**	**1990**	**1989**	**1990**	**1989**	**1990**
Import Dependence								
Imports as % of consumption	n.a	n.a	n.a	n.a	n.a	n.a	22	15
Imports as % of consumption and net exports	n.a	n.a	n.a	n.a	n.a	n.a	20	12
Share of World Consumption (%)								
Total world (approx)	n.a	n.a	n.a	n.a	n.a	n.a	n.a	n.a
Consumption Growth (% p.a.)								
1970s	n.a		n.a		n.a		-2	
1980s	n.a		n.a		n.a		-3	

REAL PRICES 1979 to 1991
Magnesium, US primary metal ingot

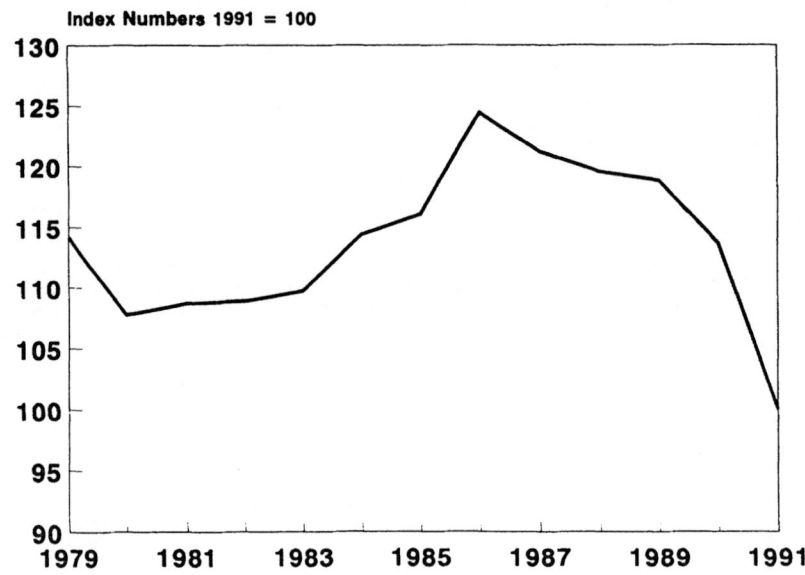

Index Numbers 1991 = 100

WORLD PRODUCTION 1979 to 1991
Magnesium Metal

Index Numbers 1991 = 100

SUPPLY AND DEMAND FOR MAGNESIUM METAL BY MAIN MARKET AREA

	UK 1989	UK 1990	EC(12) 1989	EC(12) 1990	Japan 1989	Japan 1990	USA 1989	USA 1990
Production ('000 tonnes)								
Magnesium metal primary	-	-	20.4	20.5	12.1	12.8	152.1	139.3
Magnesium metal secondary	0.6	1.0	0.6	1.0	17.8	20.4	51.2	54.5
Net Imports ('000 tonnes) Unwrought and wrought including waste and scrap	8.1	6.6	46.3	45.7	14.9	18.0	12.3	26.8
Source of Net Imports (%)								
Taiwan					3	3		
Austria			4	2				
Canada	2	5	1	2	1	2	32	76
European Community	39	37			1	1	6	4
Norway	41	32	45	41	18	25	51	14
Sweden	3	2	4	2				
Switzerland	1	2		1				
Turkey			3	3				
United States	4	12	32	35	74	62		
Yugoslavia		3	5	8				
USSR			1	2				
Others	10	7	5	4	3	7	11	6
Net Exports ('000 tonnes) Wrought & unwrought incl. waste & scrap (a) excl. France	2.4	1.5	5.6(a)	4.7(a)	0.1	0.1	56.6	51.8
Consumption ('000 tonnes)								
Incl. secondary	3.0	3.0	48.0	54.7	43.7	51.0	156.4	146.4
Import Dependence								
Imports as % of consumption	80	67	56	61	32	35	8	18
Imports as % of consumption and net exports	78	65	53	58	32	35	6	13
Share of World Consumption (%)								
Western world	1	1	16	17	14	16	51	47
Total world	1	1	12	13	11	12	38	36
Consumption Growth (% p.a.)								
1970s	-0.5		-2.0		8.5		2.8	
1980s	-5.9		0.1		3.2		1.0	

MANGANESE

WORLD RESERVES
(million tonnes manganese and % of total)

Developed			Developing			Centrally Planned		
Australia	40	(4.9)	Brazil	21	(2.6)	China	14	(1.7)
S Africa	370	(45.3)	Gabon	53	(6.5)	USSR	295	(32.2)
			Ghana	1	(0.1)			
			India	17	(2.1)			
			Mexico	4	(0.5)			
			Morocco	c.1	(0.1)			
Totals	410	(50.2)		97	(11.9)		309	(37.9)
Grand Total		816						

The reserve base is approximately 3,540 million tonnes, 96% of which is located in Australia, S Africa, Gabon and USSR. In addition, sea bed nodules contain substantial resources of manganese.

WORLD MINE PRODUCTION, 1989-90, and PRODUCTIVE CAPACITY, 1990
('000 tonnes of manganese and % of total 1990)

	Mine Production 1989	Mine Production 1990	% of Production 1990	Productive Capacity 1990
Developed				
Australia	1008	915	(10.7)	1135
S Africa	1520	1551	(18.0)	2085
Others	21	21	(0.2)	35
Total	**2549**	**2487**	**(28.9)**	**3255**
Developing				
Brazil	790	762	(8.8)	1000
Gabon	1197	1216	(14.1)	1225
Ghana	110	100	(1.2)	135
India	497	508	(5.9)	545
Mexico	150	154	(1.8)	210
Morocco	17	17	(0.2)	25
Others	30	34	(0.4	30
Total	**2791**	**2791**	**(32.4)**	**3170**
Centrally Planned				
China	635	635	7.45)	545
USSR	2722	2631	(30.6)	3000
Others	62	62	(0.7)	40
Total	**3419**	**3328**	**(38.7)**	**3585**
TOTAL	**8759**	**8606**		**10010**

The gross production of ore from which the manganese was derived was 25.1 million tonnes in 1989 and 24.7 million tonnes in 1990. The data exclude modest output of low grade ore in a number of countries.

The manganese content of mined ore varies widely between countries. The shipped ore grades of the main producers are:

	Shipped	Average
Australia	37-53	46
Brazil	30-50	38
China	20-30	20
Gabon	50-53	46(a)
Ghana	30-50	39
India	10-54	37
Mexico	27-50	38
Morocco	50-53	54
S Africa	30-48+	42
USSR	29-30	30

(a) Derived from USBM data.

World production of ferromanganese was almost 4.9 million tonnes in 1989 and 1990. About 2.5 million tonnes was produced in electric furnaces, 1.4 million tonnes in blast furnaces, and the remainder in unspecified furnaces. World production of silico manganese, where identified, totalled 1.6 million tonnes in 1989 and almost 1.5 million tonnes in 1990.

RESERVE/PRODUCTION RATIOS

Static Reserve Life (years): 95
Ratio of identified reserve base to
cumulative demand 1991-2010: 23.4 : 1 (land only)

CONSUMPTION

	'000 tonnes		% p.a. growth rates	
	1989	1990	1970s	1980s
Manganese Ore(gross weight)				
European Community	c.2600	c.2200	-0.2	-2.4
Japan	c.1900	c.1650	1.5	1.6
United States	559	497	-6.0	-7.4
Ferromanganese(gross weight)				
European Community	c.1050	c.1100	0.5	0.6
Japan	380	430	2.1	-5.9
United States	399	413	-1.6	-6.1

Note:The ferromanganese figures in this table and in the table on supply and demand by main market area may include some double counting of high carbon ferromanganese that is used to make more refined products.

MANGANESE

END USE PATTERNS, 1990 (USA)(%)

Manganese Metal

Steel (including alloy steels)	22
Super alloys	1
Aluminium alloys	67
Other alloys	8
Miscellaneous	2

Ferromanganese

Steel (including alloy steels)	95
Cast irons	3
Others (including alloys & superalloys)	2

VALUE OF ANNUAL PRODUCTION

$3.4 billion (metal content at average 1991 prices).

SUBSTITUTES

Cost and technology militate against substitution in major applications and for economic reasons there is only limited substitution in minor applications in chemicals and batteries. The steel industry has, however, made great strides in economising on the use of manganese, largely through changes in steel-making techniques.

TECHNICAL POSSIBILITIES

The mining of deep sea nodules is a potential threat to land based mines in the next century.

There is a trend towards using lower grades of ores in ferromanganese production.

New steelmaking practices and techniques are reducing the amount of manganese consumed in the process, but counterbalancing this to some extent is a trend towards higher manganese specifications in modern steels.

PRICES

	1986	1987	1988	1989	1990	1991
Ore (a)						
Europe 48-50% Mn						
$/mtu of contained						
metal	1.38	1.28	1.65	2.86	3.96	4.00
Real Dec 1991 prices	1.60	1.45	1.80	2.97	3.97	3.99
Metal (a)						
UK Electrolytic						
min 99.95%						
$/tonne	1861.9	1936.7	2036.1	2303.0	2375	2375
FERROMANGANESE						
US Imported 78% Mn						
$/tonne	314.2	337.0	502.2	609.1	634.0	553.2
Real Dec 1991 prices	364.1	380.0	546.7	632.7	636.2	551.3

(a) Source: Metal Bulletin

Prices negotiated, dependent on chemical quality, physical character, quantity, delivery terms, etc. Published quotations only reflect general condition of market. Freight charges are particularly important. Strategic value.

MARKETING ARRANGEMENTS

A few large companies dominate, with government ownership important in some cases. Five countries control the bulk of non-Eastern Bloc ore production, with South Africa as the largest supplier. There are correspondingly few companies involved. Trend to forward integration by ore producers into ferromanganese production-e.g. in South Africa. Some steel producers have manganese interests. Much ore trade is handled by agents.

REAL PRICES 1979 to 1991
Manganese, Ore 48-50% min, Europe

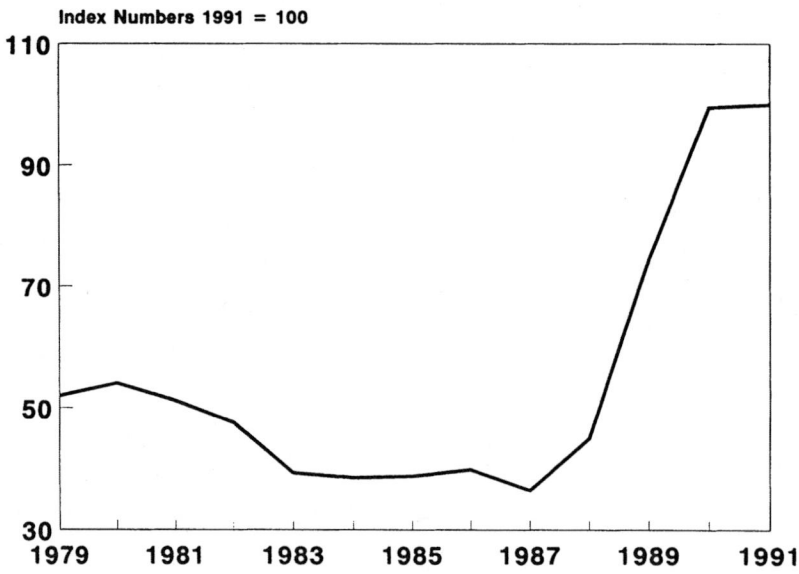

Index Numbers 1991 = 100

WORLD PRODUCTION 1979 to 1991
Manganese Ore

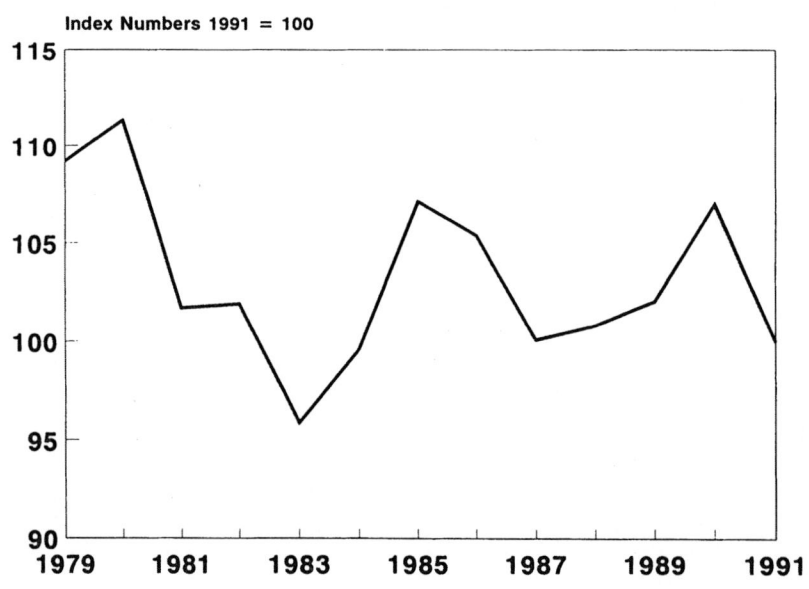

Index Numbers 1991 = 100

SUPPLY AND DEMAND BY MAIN MARKET AREA

	UK		EC(12)		Japan		USA	
	1989	1990	1989	1990	1989	1990	1989	1990
Production								
('000 tonnes)								
Mine Production								
gross weight	-	-	14.9	15	-	-		-
contained Mn	-	-	6.3	6	-	-		-
Ferromanganese	143	143	981	940	394	452	c.100	c.100
Net Imports								
('000 tonnes)								
Manganese ore gross weight	275	318	2677	2230	1939	1647	579	307
Ferromanganese gross								
weight	32.8	45.4	209	235	21.8	43.3	432	380
Metal	4.2	3.7	17.0	18.1	10.7	13.7	11.0	9.5

Source of Net Imports (%)

Manganese Ore

	UK		EC(12)		Japan		USA	
	1989	1990	1989	1990	1989	1990	1989	1990
India				3	4	4		
Australia	3	2	5	6	28	31	18	20
European Community		1	14					
S Africa	37		31	17	53	51	9	8
Brazil	57	26	15	17	8	7	34	15
Gabon			35	37	3	4	34	44
Ghana			4	4	3	2	1	2
Namibia		55		11				
Congo			7	2				
Mexico			2	1			4	2
Morocco	2	2	1	2				
Others and undefined			1		1	1		

Ferromanganese

	UK		EC(12)		Japan		USA	
	1989	1990	1989	1990	1989	1990	1989	1990
Australia							6	4
Canada					10		1	4
European Community		28	32		15	37	34	33
Japan								1
Norway	50	49	47	48	2	6	4	2
S Africa	20		45	32	38	28	38	34
Yugoslavia			1	2				1
China					22	5	2	
Brazil			1	6	2	15	3	6
India					9	3		
Mexico							8	12
USSR			4	3		6		
Namibia		17		4				
Others	1	2	2	2	2		4	1

MANGANESE

	UK 1989	UK 1990	EC(12) 1989	EC(12) 1990	Japan 1989	Japan 1990	USA 1989	USA 1990
Net Exports								
('000 tonnes)								
Manganese ore	0.8	0.4	99.0	56.8	0.1	0.1	52	70
Ferromanganese	37.9	41.2	109.0(a)	81.1(a)	15.1	5.6	8	7
Metal	2.3	1.7	2.7 (b)	1.4 (b)	1.0	0.6	5.3	6.1

(a) Exc. France
(b) Exc. UK

	UK 1989	UK 1990	EC(12) 1989	EC(12) 1990	Japan 1989	Japan 1990	USA 1989	USA 1990
Consumption								
('000 tonnes gross wt)								
Manganese ore	339	340	2600	2200	c.1900	c.1650	559	497
Ferromanganese	134.5	138.7	c.1050	c.1100	380	430	399	413

	UK 1989	UK 1990	EC(12) 1989	EC(12) 1990	Japan 1989	Japan 1990	USA 1989	USA 1990
Import Dependence (manganese)(a)								
Imports as % of consumption	100	100	99	99	100	100	100	100
Imports as % of consumption and net exports	100	100	99	99	100	100	100	100

(a)Based on mine production of manganese ore relative to consumption

	UK 1989	UK 1990	EC(12) 1989	EC(12) 1990	Japan 1989	Japan 1990	USA 1989	USA 1990
Share of World Consumption (%)								
Total world (approx.)								
Manganese ore	1	1	10	9	8	7	2	2
Ferromanganese	3	3	21	22	8	9	8	8

	UK	EC(12)	Japan	USA
Consumption Growth (% p.a.)				
1970s				
Manganese ore	-2.7	-0.2	1.5	-6.0
Ferromanganese	-4.7	0.5	2.1	-1.6
1980s				
Manganese ore	0.3	-2.4	1.6	-7.4
Ferromanganese	-0.1	0.6	-5.9	-6.1

MERCURY

WORLD RESERVES
('000 76 lb flasks and % of total)

Developed			Developing			Centrally Planned		
Spain	2200	(59.2)	Algeria	60	(1.6)	China	300	(8.1)
Turkey	100	(2.7)	Mexico	150	(4.1)	USSR	300	(8.1)
USA	100	(2.7)	Others	100	(2.7)	Others	10	(0.3)
Yugoslavia	350	(9.5)						
Others	30	(0.8)						
Totals	**2780**	**(75.1)**		**310**	**(8.4)**		**610**	**(16.5)**
Grand Total		**3700**						

The reserve base totals 7 million flasks (240,000 tonnes) with, in addition to the above, deposits in Canada and the Philippines. Identified world resources amount to 17 million flasks.

WORLD MINE PRODUCTION, 1989-90, and PRODUCTIVE CAPACITY, 1989
('000 76 lb flasks and % of total 1990)

	Mine Production		% of Production	Productive Capacity(b)
	1989	1990	1990	1989
Developed				
Finland	3.6	2.9	(1.9)	n.a
Spain	40.0	20.7	(13.8)	75
Turkey	2.9	2.9	(1.9)	8
USA	12.0	3.0	(2.0)	35
Yugoslavia	2.0	-	-	n.a
Others (a)	1.0	1.0	0.7	23
Total	**61.5**	**30.5**	**(20.3)**	**141**
Developing				
Algeria	20.3	17.4	(11.6)	35
Mexico	8.7	10.0	(6.6)	12
Total	**30.0**	**27.4**	**(18.2)**	**47**
Centrally Planned				
China	20.3	30.2	(20.1)	20
Czechoslovakia	4.8	1.4	(0.9)	n.a
USSR	60.9	60.9	(40.5)	80
Total	**86.0**	**92.5**	**(61.5)**	**100**
TOTAL	**177.5**	**150.4**		**288**

(a) Includes Finland, Yugoslavia, W Germany and Italy.
(b) Much is on standby or operating well below capacity.

MERCURY

RESERVE/PRODUCTION RATIOS

Static Reserve Life (years): 24.6
Ratio of identified reserve base to
cumulative demand 1991-2010: 2.4 : 1

CONSUMPTION

With increasingly tight environmental controls on mercury usage, demand has declined considerably in the last ten years and a growing percentage is now being met from secondary recovery. Statistics on total European demand are not available.

	'000 flasks		% p.a. growth rates	
	1989	1990	1970s	1980s
Japan	5.1	4.2	-11.5	-6.3
United States (a)	35.2	20.9	-1.4	-9.7

(a) Industrial consumption.

END USE PATTERNS (%)

USA(1989)

Batteries	21
Electrical apparatus	14
Mildew proofing paint	6
Electrolytic production of chlorine/caustic soda	31
Laboratory, catalysts & chemicals	5
Instruments	7
Dentistry	3
Others	3

Japan(1990)

Batteries	65
Chemicals	12
Others	23

VALUE OF CONTAINED METAL IN ANNUAL PRODUCTION

$19 million (at average 1991 prices).

SUBSTITUTES

Lithium and nickel-cadmium batteries are increasingly used alternatives for mercury batteries but, generally, there are few satisfactory substitutes for applications in electrical apparatus and industrial and control instruments.

Diaphragm and membrane cells are rapidly replacing cells using mercury in the chlor-alkali industry.

Organic mildewicides are being substituted in latex paints; plastic paint and copper oxide paint are being used to protect ship hulls.

TECHNICAL POSSIBILITIES

Environmental considerations are encouraging conservation and recycling.

Design changes in mercury cell and improvements in diaphragm cell are modifying consumption.

PRICES

	1986	1987	1988	1989	1990	1991
New York Dealer Price 99.99% $/flask of 76 lb 20+ flask lots	239.2	300.5	342.9	294.5	257.0	127.6
Real Dec 1991 prices	277.7	339.6	374.1	306.3	258.8	127.0

Until 1978/79, markets were dominated by dealers but producer pricing has become more important since then, particularly outside the USA. Large quantities of secondary material and exports from China and the USSR affect prices.

MARKETING ARRANGEMENTS

The major producers' hold on the market has weakened as increased volumes of secondary material have become available plus supplies from China, the USSR, and the US stockpile. ASSIMER, the Mercury Producers' Association, whose members include Spain, Italy, Yugoslavia and Algeria, has had a varying impact on the market. Over three-quarters of world production is from state owned or controlled mines. The largest producer, Minas de Almaden y Arrayanes of Spain, forced up prices in 1987/88, but prices have since slumped because of weakening markets caused by a combination of recession and environmental controls. Mines have been forced to curtail production or close down.

REAL PRICES 1979 to 1991
Mercury, New York Dealer

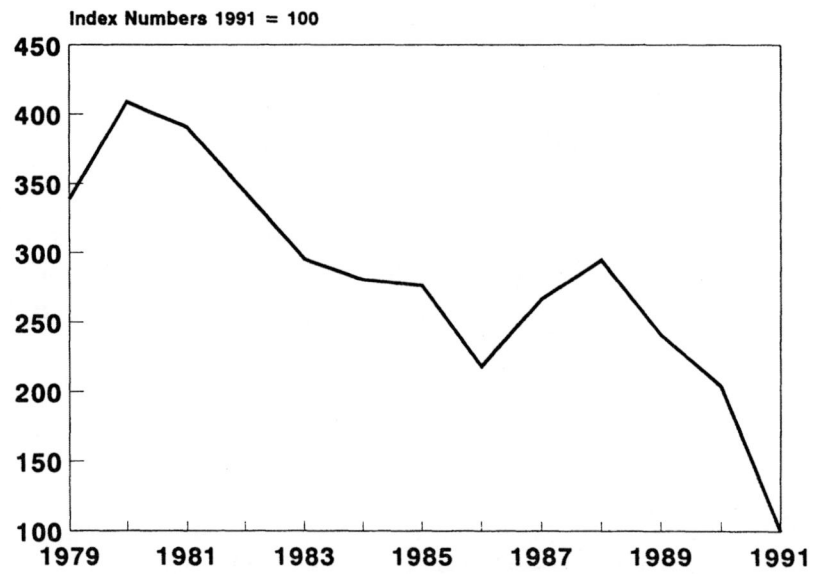

Index Numbers 1991 = 100

WORLD PRODUCTION 1979 to 1991
Mercury

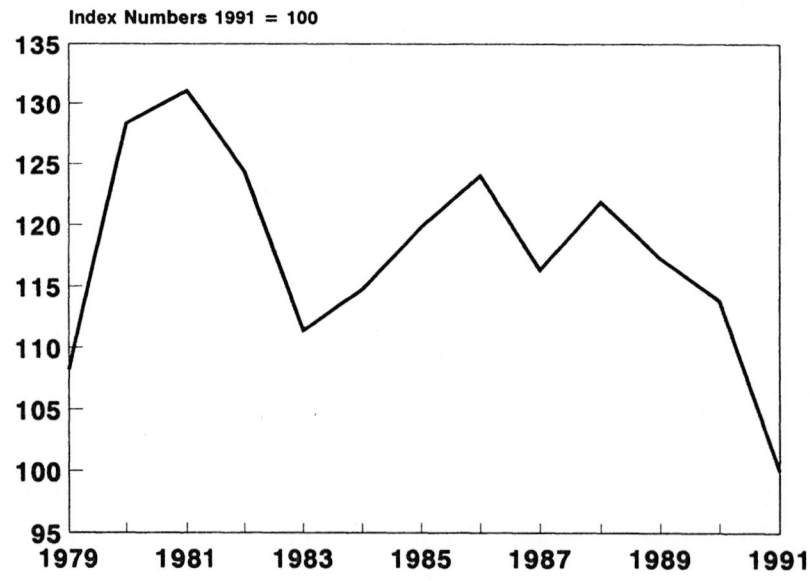

Index Numbers 1991 = 100

SUPPLY AND DEMAND BY MAIN MARKET AREA

	UK		EC(12)		Japan		USA	
	1989	1990	1989	1990	1989	1990	1989	1990
Production								
('000 76 lb flasks)								
Primary	-	-	40	20.7	-	-	12.0	3.0
Secondary	n.a	n.a	n.a	n.a	n.a	n.a	4.0	2.2
NDS releases	-	-	-	-	-	-	4.9	1.0
US DoE releases	-	-	-	-	-	-	5.2	2.7
Net Imports								
('000 76 lb flasks)	0.09	0.05	0.45	0.34	1.80	1.00	3.80	0.43
Source of Net Imports (%)								
European Community	89	79			30	13	97	43
Taiwan					19	8		
Japan								
Turkey	5		25					
Canada								57
USA								
Algeria	1		20		11	10		
China					40	69		
Hong Kong								
Others and unidentified	5	21	55	n.a				3
Net Exports								
('000 76 lb flasks)	0.12	0.18	0.44	0.26	5.9	3.0	6.4	9.03
						(mainly secondary from stock)		
Consumption								
('000 76 lb flasks)	n.a	n.a	n.a	n.a	5.1	4.2	35.2	20.9
		(primary only)				(apparent)		(industrial)
Import Dependence								
Imports as % of consumption	100	100	n.a	n.a	35	24	11	2
		(primary only)						
Imports as % of consumption and net exports	100	100	n.a	n.a	16	14	9	1
		(primary only)						

A large percentage of European demand is met from stocks. This complicates the calculation of import shares.

Share of World Consumption(%)
Because of the substantial tonnages of mercury consumed from secondary recovery or from stocks, and the lack of complete statistics thereon, reliable estimates of shares of world consumption cannot be made.

Consumption Growth (% p.a.)

1970s	9.7	declined	-11.5	-1.4
	(primary only)	rapidly		
1980s	n.a	n.a	-6.3	-9.7

MOLYBDENUM

WORLD RESERVES
('000 tonnes of metal and % of total)

Developed			Developing			Centrally Planned		
Canada	450	(8.1)	Chile	1130	(20.5)	China	500	(9.1)
USA	2720	(49.1)	Iran	50	(0.9)	USSR	450	(8.1)
			Mexico	90	(1.6)			
			Peru	140	(2.5)			
Totals	3170	(57.2)		1410	(25.5)		950	(17.2)
Grand Total	5530							

The world reserve base is 11.8 million tonnes mainly located in the USA, Canada, Chile, the USSR and China. Identified resources amount to approximately 17 million tonnes.

WORLD MINE PRODUCTION, 1989-90, and PRODUCTIVE CAPACITY, 1990
('000 tonnes of metal and % of total 1990)

	Mine Production		% of Production	Productive Capacity
	1989	1990	1990	1990
Developed				
Canada	13.7	13.6	12.4	15.9
Japan	0.1	0.1	0.1	-
USA	63.1	61.2	55.8	74.8
Total	**76.9**	**74.9**	**68.3**	**90.7**
Developing				
Chile	16.6	13.8	12.6	20.4
India	0.1	0.1	(0.1)	n.a
Iran	0.5	0.5	0.5	1.8
Mexico	4.2	3.2	2.9	6.8
Niger	0.1	0.1	0.1	n.a
Peru	3.2	2.5	2.3	4.5
S Korea	0.1	0.1	0.1	0.9
Turkey	0.1	0.1	(0.1)	n.a
Total	**24.9**	**20.4**	**(18.6)**	**34.4**
Centrally Planned				
Bulgaria	0.2	0.2	0.2	0.9
China	2.0	2.0	1.8	4.5
Mongolia	1.2	1.2	1.1	2.3
USSR	11.5	11.0	10.0	15.9
Total	**14.9**	**14.4**	**(13.1)**	**25.6**
TOTAL	**116.7**	**109.7**		**148.7**

A number of other countries, including N Korea and Turkey, produce molybdenum but no reliable data are available to determine production.

RESERVE/PRODUCTION RATIOS

Static Reserve Life (years):	50
Ratio of identified reserve base to cumulative demand 1991-2010:	6.1 : 1

CONSUMPTION
(Molybdenum in all forms)

	tonnes		% p.a. growth rates	
	1989	1990	1970s	1980s(a)
European Community	29500	28425	2.3	1.0
Japan	14250	16220	4.4	3.2
United States	23040	20500	3.1	-2.4
Other Countries	17985	17045	2.4	5.9
Total Western World	**84775**	**82190**	**2.5**	**1.1**

(excl. exports to Eastern Countries)

Source:International Molybdenum Association and CRU
(a) Base don Climax Molybdenum's data

END USE PATTERNS, 1990 (%)

USA

Steel	56
Cast irons	4
Super and special alloys	12
Molybdenum metal	11
Chemicals and ceramic use	9
Others	8

Western World

Construction Steel	33
Stainless Steel	28
Tool Steel	8
Cast Iron and steel	6
Metal & super alloys	13
Chemicals	13

VALUE OF CONTAINED METAL IN ANNUAL PRODUCTION

$0.58 billion (at average 1991 prices).

SUBSTITUTES

Potential substitutes in alloy steel include boron, chromium, manganese, columbium, vanadium and nickel. Tungsten can be used in tool steels and along with tantalum, in certain refractory metal uses. Graphite can replace molybdenum for refractory elements in some electric furnaces. Chrome orange, cadmium red and organic orange pigments are substitutes for molybdenum orange. Most of the above alternatives to molybdenum suffer losses in efficiency. Heat treatment of alloy steels is an alternative to molybdenum.

TECHNICAL POSSIBILITIES

Increased molybdenum recovery through improvement in efficiency of flotation techniques.

Development and application of new molybdenum-based steels and alloys particularly of resistance to oxidation at high temperatures is improved.

MOLYBDENUM

PRICES

	1986	1987	1988	1989	1990	1991
By-product Concentrate 95% MoS_2 $/1b	2.49	2.59	2.66	3.18	2.57	2.13
Dealer Oxide molybdic trioxide, $/lb						
Average	2.92	2.95	3.47	3.39	2.81	2.35
Real Dec 1991 prices	3.37	3.33	3.79	3.53	2.81	2.34

Prices for ferromolybdenum are linked to concentrate prices. Prior to the 1979-80 upsurge in prices molybdenum was mainly producer priced, with a dealer market that influenced producer price movements. The dealer market subsequently became much more important, and producers' effective prices followed the market. By-product material was normally sold at discounts from the Climax price. Production cutbacks in early 1980s temporarily forced concentrate to a premium over oxide causing problems for independent roasters. Producer prices were reinstated by Cyprus Mines and Amax in 1986, and since met with varying success. Recession in 1990-91 forced down prices and prompted a renewed round of closures by primary producers attempting to prevent prices from falling.

MARKETING ARRANGEMENTS

Under 30 mines in USA, Canada, Chile, Mexico and Peru account for most of the world's production. Cyprus, Amax and Codelco the Chilean copper producer. Noranda and Placer Dome have painfully established tenuous control over the markets. They purchase most of the by-product concentrate, as well as sell from their own mines.

The International Molybdenum Association is a grouping of most major producers and intermediate processors for statistical and research purposes.

REAL PRICES 1979 to 1991
Molybdenum, Dealer Oxide

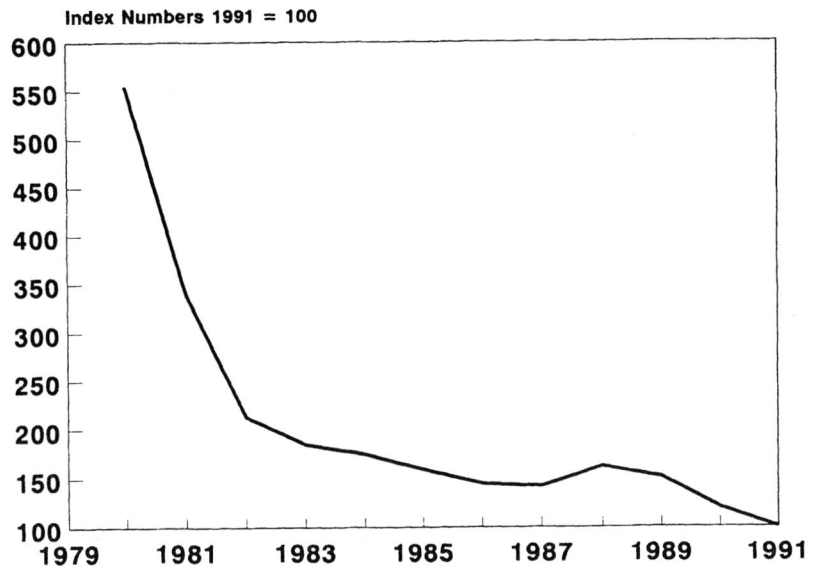

Index Numbers 1991 = 100

WORLD PRODUCTION 1979 to 1991
Molybdenum

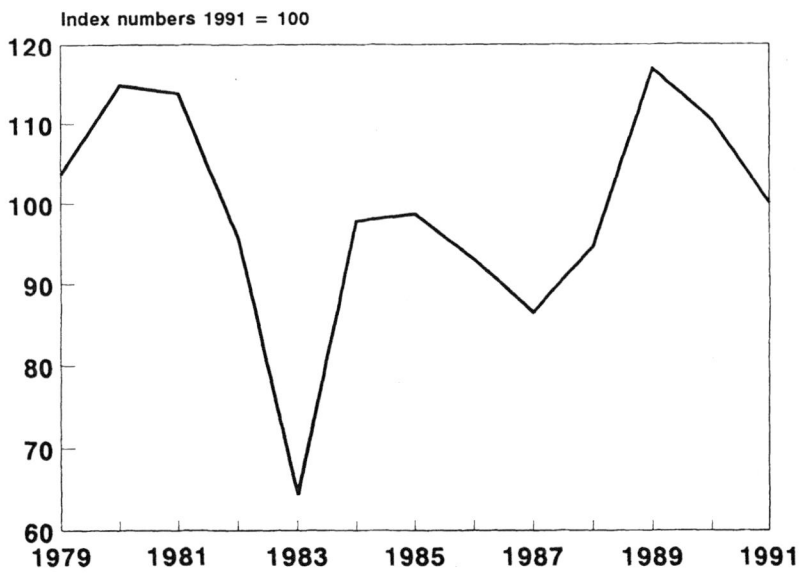

Index numbers 1991 = 100

MOLYBDENUM

SUPPLY AND DEMAND IN MAIN MARKET AREAS

	UK		EC(12)		Japan		USA	
	1989	**1990**	**1989**	**1990**	**1989**	**1990**	**1989**	**1990**
Production (tonnes)								
Mine production (Mo content)	-	-	-	-	-	-	63105	61611
Net Imports (tonnes)								
Ores and concentrates	22990	18115	89976	81561	23106	24408	619 258(Mo)	1054 479(Mo)
Ferromolybdenum	635	502	3268	4945	1322	1600	1410	1415
Oxides	146	160	261	211	352	335	422	669
Metal (unwrought)	10	19	150	253	295(a)	198(a)	60	58
Powders	26	49	64	189	160	150	216	79
All forms (Mo)	-	-	-	-	-	-	2823	2497

(a)Includes waste and scrap

Source of Net Imports (%)

	UK		EC(12)		Japan		USA	
Ores and concentrates								
Canada	1	4	13	13	30	33	2	
European Community		12	20			6	11	5
Sweden	1		1	2				
United States	79	73	67	65	26	25		
China	2		2	1				1
Chile	4	2	13	13	35	31		3
Mexico		1	2	3	3		93	95
Peru	1		2	3				
Others								1

	UK		EC(12)		Japan		USA	
Net Exports (tonnes)								
Ores and concentrates	2465	3060	17010	16476	-	-	5123 (Mo)	19890 (Mo)
Ferromolybdenum	6696	5282	2733(a)	1306(a)	2	2	75	300
Metal(unwrought)	68	37	78(b)	72(b)	5(c)	10(c)	253	180
Oxides	661	557	1370	1592	29	9	1391	787
Powders	404	269	152(b)	208(b)	8	7	632	292

(a) Excludes Belgium-Luxembourg
(b) Excl. Germany
(c) Includes waste & scrap

	UK		EC(12)		Japan		USA	
Consumption (tonnes)								
All forms (Mo content)	7800 3130 (steel use)	5200 2960	29500	28425	14250	16220	23040	20500

	UK		EC(12)		Japan		USA	
	1989	1990	1989	1990	1989	1990	1989	1990
Import Dependence								
Imports as % of consumption	100	100	100	100	100	100	-	-
Imports as % of consumption and net exports	100	100	100	100	100	100	-	-
Share of World Consumption (%)								
Western World	9	6	35	35	17	20	27	25
Consumption Growth (% p.a.)								
1970s	-3.5		2.3		4.4		3.1	
1980s	-0.2		1.0		3.2		-2.4	

NICKEL

WORLD RESERVES
('000 tonnes of contained nickel and % of total)

Developed			Developing			Centrally Planned		
Australia	2180	(4.6)	Botswana	475	(1.0)	Albania	180	(0.4)
Canada	6170	(12.9)	Brazil	665	(1.4)	China	725	(1.5)
Finland	80	(0.2)	Colombia	560	(1.2)	Cuba	18145	(38.1)
Greece	455	(1.0)	Dominican Rep.	455	(1.0)	USSR	6620	(13.9)
S Africa	2540	(5.3)	Indonesia	3200	(6.7)			
USA	25	(0.1)	New Caledonia	4535	(9.5)			
Yugoslavia	155	(0.3)	Philippines	410	(0.9)			
			Zimbabwe	75	(0.2)			
Totals	**11605**	**(24.4)**		**10375**	**(21.8)**		**25670**	**(53.9)**
Grand Total		**47650**						

The world's reserve base is estimated at 110 million tonnes and, in addition to the above countries, includes deposits in Guatemala, Papua New Guinea and several African nations. The average grade of the reserves included exceeds 1% nickel.

Identified world resources of nickel in deposits averaging 1% nickel or more contain 130 million tonnes of which 60% is in laterites. Resources of lower grade deposits are very large, and there are extensive sea bed resources of nickel in manganese nodules.

WORLD MINE PRODUCTION, 1989-90, and PRODUCTIVE CAPACITY, 1990
('000 tonnes of nickel and % of total 1990)

	Mine Production		% of Production	Productive Capacity
	1989	1990	1990	
Developed				
Australia	65.0	67.0	(7.7)	75
Canada	200.9	199.4	(22.8)	200
Finland	10.5	11.5	(1.3)	11
Greece	18.9	18.5	(2.1)	23
Norway	1.3	3.1	(0.4)	-
S Africa	34.0	30.0	(3.4)	45
Yugoslavia	6.3	4.9	(0.6)	n.a*
USA	-	0.3	-	4
Total	**336.9**	**334.7**	**(38.3)**	**358**
Developing				
Botswana	19.8	19.0	(2.2)	20
Brazil	13.7	13.4	(1.5)	24
Myannar	0.1	0.1	(-)	-
Colombia	16.9	18.4	(2.1)	22
Dominican Rep.	31.3	28.7	(3.3)	32
Indonesia	59.6	53.8	(6.2)	64
New Caledonia	80.3	85.1	(9.7)	91
Philippines	15.4	15.8	(1.8)	41(b)
Zimbabwe	12.7	12.6	(1.4)	18
Total	**249.8**	**246.9**	**(28.2)**	**312**
Centrally Planned				
Albania	8.8	8.5	(1.0)	10
China	26.0	28.0	(3.2)	36
Cuba	46.5	43.2	(4.9)	55
USSR	205.0	212.0	(24.3)	304
Others	1.7	1.0	(0.1)	n.a*
Total	**288.0**	**292.7**	**(33.5)**	**405**
Not specified(a)				27
TOTAL	**874.7**	**874.3**	**100.0**	**1102**

*Included in 'not specified'.
(a) Austria, Czechoslovakia, Germany, Poland & Yugoslavia
(b) Largely on standby.

NICKEL

WORLD REFINED METAL PRODUCTION, 1989-90
('000 tonnes of nickel and % of total 1990)

Developed	1989	1990	% 1990	Developing	1989	1990	% 1990	Centrally Planned	1989	1990	% 1990
Australia	42.9	46.7	(5.4)	Brazil	13.7	13.4	(1.6)	Albania	5.5	4.0	(0.5)
Austria	2.8	3.3	(0.4)	Colombia	16.9	18.4	(2.1)	Cuba	26.5	21.0	(2.4)
Canada	129.1	126.8	(14.7)	Dominican Rep.	31.3	28.7	(3.3)	USSR	225.0	230.0	(26.7)
France	8.6	8.5	(1.0)	Indonesia	5.0	5.0	(0.6)	E Germany	2.7	1.3	(0.2)
Finland	13.4	16.9	(2.0)	New Caledonia	36.3	32.3	(3.7)	Czecho-			
Greece	9.2	13.1	(1.6)	Taiwan	10.0	10.4	(1.2)	slovakia	2.5	2.6	(0.3)
Japan	105.8	100.3	(11.6)	Zimbabwe	18.8	18.8	(2.2)	China	26.3	27.5	(3.2)
Norway	54.9	57.8	(6.7)	S. Korea	5.0	8.2	(1.0)				
S Africa	30.0	28.0	(3.3)								
UK	26.1	26.5	(3.1)								
Yugoslavia	6.3	4.9	(0.6)								
USA	0.3	3.7	(0.4)								
Totals	436.3	439.1	(51.0)		137.0	135.2	(15.7)		288.5	286.4	(33.3)
Grand Totals	1989-	861.8									
	1990-	860.7									

RESERVE/PRODUCTION RATIO

Static Reserve Life (years):	55
Ratio of identified reserve base to cumulative demand 1991-2010:	4.8 : 1 (land based only)

CONSUMPTION

	'000 tonnes		% p.a. growth rates		
	1989	1990	1960s	1970s	1980s
European Community	226.2	238.5	5.9	3.4	3.5
Japan	163.0	164.2	18.9	4.3	2.9
United States	120.2	127.9	3.2	1.8	-1.1
Others	148.4	147.1	11.0	5.7	5.3
Total Western World	**657.8**	**677.7**	**7.0**	**3.4**	**2.6**
Total World	**847.4**	**849.9**	**6.9**	**3.6**	**1.9**

END USE PATTERNS, 1990 (%)

	USA	Japan		USA
Stainless and alloy steels	37	71	Transport	26
Non-ferrous alloys	30	8	Chemical industry	12
Electroplating	15	9	Electrical equipment	8
Others	18	12	Construction	8
			Fabricated metal products	7
			Petroleum	5
			Household appliances	6
			Machinery	6
			Other	22

VALUE OF CONTAINED METAL IN ANNUAL PRODUCTION

$7.0 billion (refined metal at 1991 average LME prices).

SUBSTITUTES

The use of alternative materials tends to be more expensive or requires some sacrifice in chemical or physical characteristics, and hence performance. Alternative materials are however available to replace nickel in most of its uses. Alloy substitutes are normally other 'steel' industry metals such as molybdenum, columbium and manganese. Platinum, cobalt and copper can be used in some catalysts. Titanium and many plastics can compete for markets where corrosion-resistance is important. Cobalt can replace nickel in electroplating applications.

TECHNICAL POSSIBILITIES

Deep sea nodules.

Development of new nickel-bearing alloys.

Substitution of nickel-based superalloys by ceramic components.
Development of nickel metal-hydride batteries.

PRICES

	1986	1987	1988	1989	1990	1991
Cathode						
US Dealer $/lb	1.88	2.29	6.24	6.14	4.07	3.80
LME Cash £/tonne	2645	2954	7726	-	-	-
LME Cash US$/tonne	-	-	-	8046	8864.1	8155.6
LME Cash US $/lb	1.76	2.22	6.26	6.03	4.02	3.70
Real Dec 1991 prices	2.04	2.51	6.83	6.28	4.02	3.69

Producer pricing gave way to a predominantly dealer market after a London Metal Exchange quotation was introduced in mid-1979. The majority of nickel is still traded on producer-consumer contracts, but at LME-related prices. Breakeven costs are influenced by associated by-product revenues.

MARKETING ARRANGEMENTS

The influence of major producers has weakened, although International Nickel (Inco) still retains over one-third of the Western world market, with Falconbridge, SLN and Western Mining as other major producers. State participation in the industry is increasing, mainly through joint ventures. Dealer markets, including the LME, backed by substantial Western imports from Cuba and the USSR, have dominated pricing in recent years. All major producers have managed to reduce their costs, sometimes substantially. Prices rose sharply in 1988 on the back of strong demand from the stainless steel industry and a lack of available capacity, but subsided once balance was restored.

An intergovernmental International Nickel Study Group, based in The Hague, was established in 1991, under UN auspices. With membership drawn from producer and consumer countries, it produces statistics and discusses matters of interest to the nickel industry.

REAL PRICES 1979 to 1991
Nickel, LME Cash

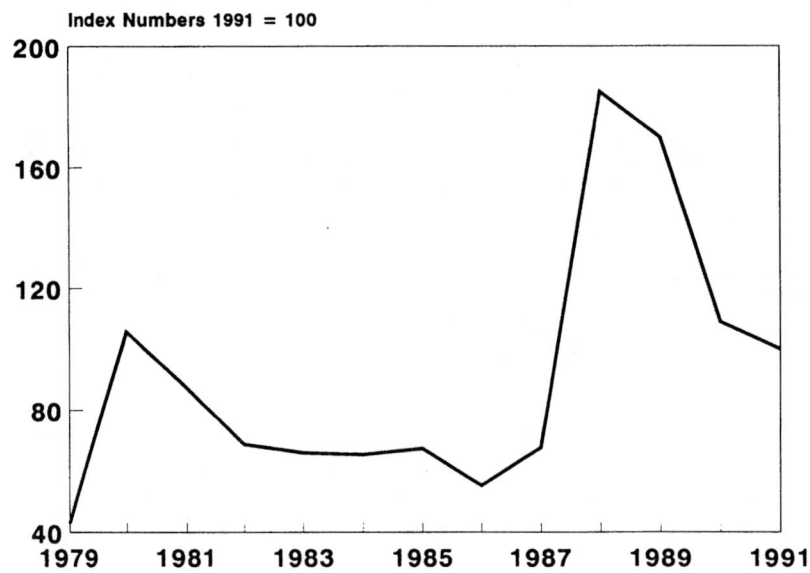

Index Numbers 1991 = 100

WORLD PRODUCTION 1979 to 1991
Nickel

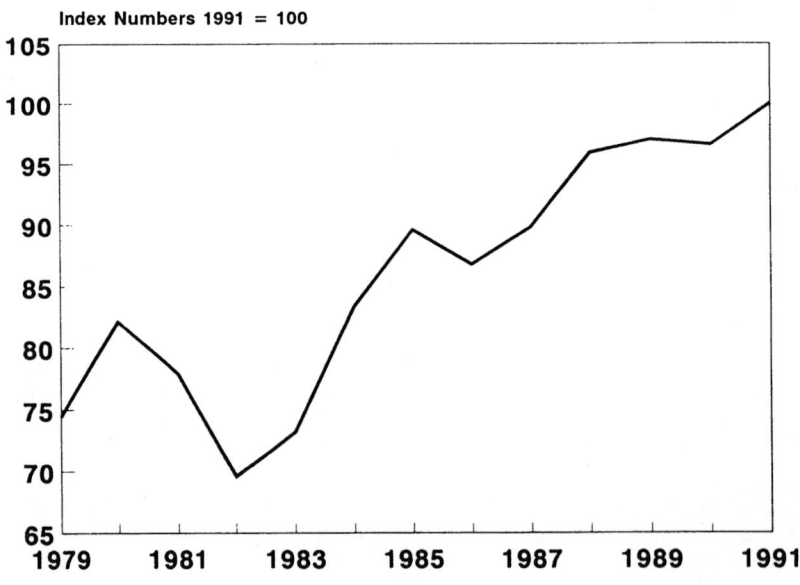

Index Numbers 1991 = 100

SUPPLY AND DEMAND BY MAIN MARKET AREA

	UK 1989	UK 1990	EC(12) 1989	EC(12) 1990	Japan 1989	Japan 1990	USA 1989	USA 1990
Production								
('000 tonnes Ni content)								
Mine production	-	-	18.9	18.5	-	-	-	0.31
Smelter/Refinery								
Production	26.1	26.5	50.8	50.7	105.8	100.3	0.3	3.7
of which								
Metal	26.1	26.5	34.7	35.8	21.9	22.3	-	-
Ferro and Nickel								
oxide sinter	-	-	16.1	14.9	83.9	78.0	0.3	3.7
Net Imports								
('000 tonnes Ni content)								
Ores and concentrates	-	-	-	-	61.4	67.0	-	-
Ferro Nickel	5.3	4.7	45.7	42.0	14.1	11.4	11.5	14.3
Nickel Matte	25.2	32.9	36.0	43.5	44.1	39.4	-	-
Oxide/Sinter	-	4.2	20.6	24.8	2.2	2.2	0.1	0.9
Unwrought Nickel	17.1	18.0	118.7	119.7	38.0	44.0	93.4	104.5
Source of Net Imports(%)								
Ferro Nickel								
New Caledonia		8	41	34	44	46	38	19
Indonesia	8	10	4	8	20	25	2	
Dominican Republic	24	7	27	19	26	16	45	60
Columbia	5	2	10	18	10	12	14	20
European Community		47	60					
Canada	9	10	1					
Austria	6		3	2				
Others	1	3	14	19		1	1	1
Nickel Matte								
Indonesia					61	63		
Australia					39	37		
New Caledonia				30	24			
Canada	97	100	70	76				
USA	3							
Unwrought Nickel								
USSR	24	23	45	45	34	38		
Zimbabwe	4	4	4	3	15	15	3	3
Norway	10	12	14	14	18	15	16	21
Canada	19	12	13	10	14	12	69	61
Australia	18	15	7	9	5	5	6	7
Finland	5	9	3	4	3	2	1	1
European Community	2	3		7	8	2	2	
Taiwan				1				
South Africa	15	18	10	11	2	2	1	2
Dominican Republic							2	
Others	3	4	4	4	1	3		3

NICKEL

	UK 1989	UK 1990	EC(12) 1989	EC(12) 1990	Japan 1989	Japan 1990	USA 1989	USA 1990
Net Exports								
('000 tonnes Ni content)								
Ferro Nickel	-	-	1.9	5.0	-	-	0.2	1.7
Nickel Matte	-	-	-	-	-	-	-	-
Oxide/Sinter	-	0.2	-	0.2	-	-	-	-
Nickel	11.8	14.3	8.5	9.3	0.1	0.1	1.1	1.5
Consumption								
('000 tonnes Ni content)								
All forms	29.5	32.6	226.2	238.3	163.0	164.2	120.2	127.9
Import Dependence								
Imports as % of consumption	100	100	92	92	100	100	100	100
Imports as % of consumption and net exports	100	100	92	92	100	100	100	100
Share of World Consumption (%)								
Western world	5	5	34	35	25	24	18	19
Total world	4	4	27	28	19	19	14	15
Consumption Growth (% p.a.)								
1960s	1.6		5.9		18.9		3.2	
1970s	-0.3		3.4		4.3		1.8	
1980s	3.6		3.5		2.9		-1.1	

NIOBIUM

WORLD RESERVES
('000 tonnes of metal and % of total)

Developed			Developing			Centrally Planned		
Canada	136	(3.2)	Brazil	3310	(78.3)	USSR	c.680	(16.1)
			Nigeria	65	(1.5)			
			Malaysia					
			Thailand }	6	(0.1)			
			Zaire	32	(0.8)			
Totals	136	(3.2)		3413	(80.7)		c.680	(16.1)
Grand Total		4229						

The reserve base is approximately 5 million tonnes.

WESTERN WORLD MINE PRODUCTION, 1989-90, and PRODUCTIVE CAPACITY, 1990
(tonnes of contained metal and % of total 1990)

	Mine Production		% of Production	Productive Capacity
	1989	1990	1990	1990
Developed				
Australia	64	64	(0.5)	90
Canada	2358	2458	(17.9)	2500
Total	**2422**	**2522**	**(18.4)**	**2590**
Developing				
Brazil	14295	11159	(81.3)	18145
Malaysia	-	-	(..)	90
Namibia	..	1	(..)	(a)
Nigeria	20	19	(0.1)	180
Thailand	19	1	(..)	270
Zaire	13	13	(0.1)	45
Zimbabwe	5	5	(0.1)	90(a)
Total	**14352**	**11198**	**(81.6)**	**18820**
TOTAL	**16774**	**13720**		**21410**

(a) Includes Mozambique, Rwanda and Namibia.

The above data are mostly US Bureau of Mines estimates based on the reported gross weight of production. Spain, Zambia, the USSR and China also produce, or are thought to produce, niobium but reliable estimates of output are not available.

Pyrochlore, the main ore of Brazil and Canada, contains about 42% niobium. The content of columbite-tantalite ores is much lower, but highly variable. The Tantalum-Niobium International Study Centre estimates 1989 Western World mine production at 14056 tonnes contained niobium, and 1990 output at 15340 tonnes.

NIOBIUM

RESERVE/PRODUCTION RATIOS

Static Reserve Life (years):   over 300
Ratio of identified reserve base to
cumulative demand 1991-2010:   1.1 : 1

CONSUMPTION

Reliable statistics are not available for most areas but broad orders of magnitude are as follows for contained niobium in all forms.

	tonnes		% p.a. growth rates	
	1989	1990	1970s	1980s
European Community	c.3200	c.2900	approx 5 to 8	c.-0.3
Japan	c.2650	c.2750	12.1	c.3.6
United States (a)	3400	3360	4.1	-0.3

(a) Apparent consumption.
Ferroniobium contains about 65% niobium

END USE PATTERNS, 1990 (%)

USA

HSLA Steels	31
Carbon steels	36
Stainless + heat-resisting steels	14
Superalloys	18
Others	1

Western World

HSLA grade ferroniobium	87
Metal & alloys	2
Compounds & additives	11

VALUE OF CONTAINED METAL IN ANNUAL PRODUCTION

$120 million for niobium contained in ore (Western World only at average 1991 prices).

SUBSTITUTES

Substitutes usually lower performance and/or cost effectiveness.

Vanadium, titanium and molybdenum in HSLA steels. Tantalum competes in superalloys. Titanium can be used in stainless steels. In high temperature applications, molybdenum, tungsten, tantalum and ceramics are alternatives.

TECHNICAL POSSIBILITIES

Refinements in beneficiating and processing techniques are giving products of higher purity or different composition.

There is continuing development of new steels, superalloys, superconductors for low-temperature usage and super-conducting magnets.

PRICES

	1986	1987	1988	1989	1990	1991
Ore						
Canadian Pyrochlore US cents/lb contained Nb_2O_5	260	260	260	260	260(a)	260(a)
Real Dec 1991 prices	301.7	293.8	283.8	270.4	260.9	259.1

(a) Nominal prices. Prices for spot lots of columbite ore (Metal Bulletin) were $2.5-2.9/lb contained Nb_2O_5 up to April 1990 and $2.6-3.05/lb subsequently.

Mainly producer price basis and nominal price changes are infrequent. Concentrate producers have low costs relative to prices. Outside of Canada and Brazil, most niobium is produced as a by-product of tin mining.

MARKETING ARRANGEMENTS

The Araxa mine (Companhia Brasilira de Metallurgica e Mineracao) and the Catala mine (Mineracao Catalao de Goias Ltda) in Brazil, and Niobec in Canada are the major concentrate producers, and dominate the market. Most of the Brazilian material is processed into ferroniobium before export. Production of metal is usually in the hands of separate concerns from the mining companies.

REAL PRICES 1979 to 1991
Niobium, Canadian Pyrochlore

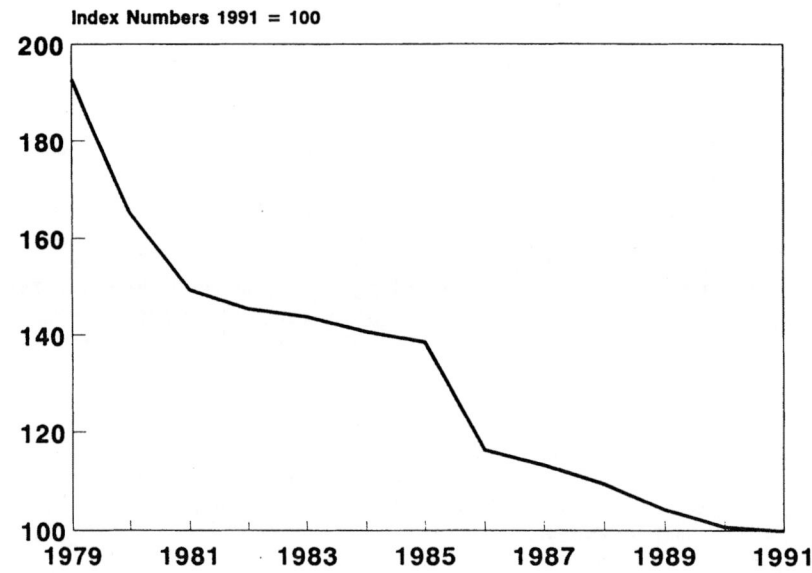

Index Numbers 1991 = 100

WESTERN WORLD PRODUCTION 1979 to 1991
Niobium

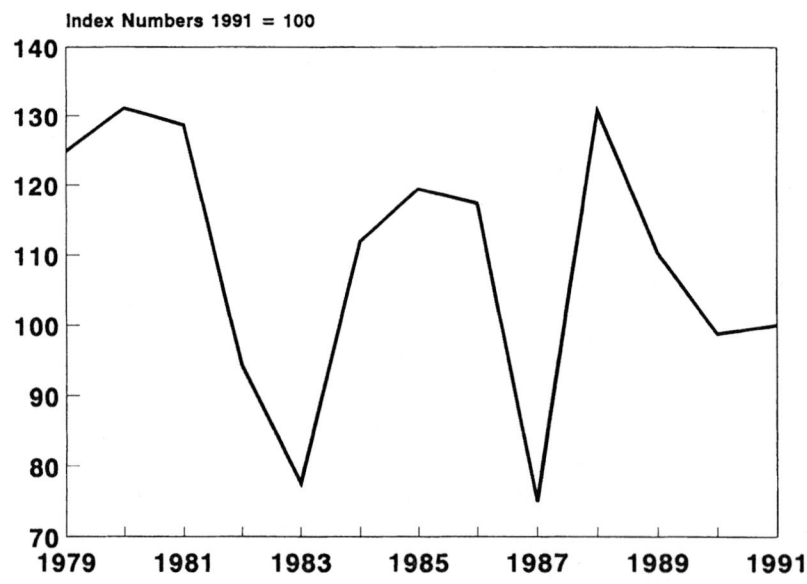

Index Numbers 1991 = 100

SUPPLY AND DEMAND BY MAIN MARKET AREA

	UK		EC(12)		Japan		USA	
	1989	1990	1989	1990	1989	1990	1989	1990
Production								
(tonnes)								
Mine production	-	-	n.a	n.a	-	-	-	-
Ferroniobium:								
approx. gross wt	n.a	n.a	n.a	n.a	737	c.750	n.a	n.a
Net Imports								
(tonnes)								
Ores and Concentrates								
(gross wt)	2270(a)	1432(a)	2270(a)	1445(a)	816	1171	2227	1950
Ferroniobium (gross wt)	459	472	4241	4035	3590	3516	3880	3013
Metal and alloys	27(b)	25(b)	41(b)	55(b)	n.a	n.a	9	2
				(incl. wrought &	(incl. with gallium)			
				waste & scrap)				

(a) Includes tantalum ores and concentrates.
(b) Includes rhenium.

Sources of Net Imports(%)

Ores and Concentrates								
Canada	95	99	95	98	91	81	99	99
European Community								
United States								
Brazil					4	9	1	1
Nigeria					5	9		
Others	5	1	5	2		1		
Ferroniobium								
European Community	9						8	2
Brazil	91		100		100	100	92	98
Other								

Net Exports

(tonnes)								
Ores & Concentrates	6(a)	3(a)	46(a)	3(a)	-	-	127	64
Ferroniobium(gross wt)	804	812	257	287	-	-	168	838
Metal compounds								
and alloys	1	3(b)	6(b)	17(b)	Incl. with gallium			

(a) Excluding UK.
(b) Includes rhenium.

NIOBIUM

	UK 1989	UK 1990	EC(12) 1989	EC(12) 1990	Japan 1989	Japan 1990	USA 1989	USA 1990
Consumption (tonnes)								
Niobium metal	n.a	n.a	n.a	n.a	n.a	n.a	n.a	n.a
Ferroniobium (Nb content)	490	430	n.a	n.a	n.a	n.a	2439	2585
	(iron & steel usage)							
Total all forms	c.500	c.450	c.3200	c.2900	c.2650	c.2750	3400	3360
Import Dependence								
Imports as % of consumption	100	100	100	100	100	100	100	100
Imports as % of consumption and net exports	100	100	100	100	100	100	100	100
Share of World Consumption (%)								
Westerm World (approx.)	2	3	19	21	16	20	20	25
Consumption Growth (% p.a.)			approx.					
1970s	-5		5 to 8		12.1		4.1	
			(ferro only)					
1980s	-		c.-0.3		c.3.6		-0.3	

PHOSPHATE

WORLD RESERVES
(million tonnes and % of total)

Developed			Developing			Centrally Planned		
S Africa	2530	(20.2)	Brazil	33	(0.2)	China	210	(1.6)
USA	1230	(9.7)	Christmas Is	10	(0.1)	USSR	1330	(10.6)
Others	450	(3.6)	Jordan	90	(0.7)	Others	237	(1.9)
			Morocco W Sahara }	5900	(46.9)			
			Nauru	5	(..)			
			Others	560	(4.5)			
Totals	4210	(33.5)		6598	(52.4)		1777	(14.1)
Grand Total		12585						

This table uses production costs of under $40/tonne.

WORLD RESERVE BASE
(million tonnes and % of total)

Developed			Developing			Centrally Planned		
Australia	590	(1.7)	Algeria	240	(0.7)	China	210	(0.6)
Canada	50	(0.1)	Brazil	370	(1.1)	USSR	1330	(3.9)
Finland	70	(0.2)	Christmas Is	10	(..)	Others	(..)	(..)
S Africa	2530	(7.4)	Colombia	100	(0.3)			
Turkey	30	(0.1)	Egypt	760	(2.2)			
USA	4440	(13.0)	Israel	180	(0.5)			
			Jordan	480	(1.4)			
			Mexico	110	(0.3)			
			Morocco W Sahara }	21440	(62.6)			
			Nauru	5	(..)			
			Senegal	160	(0.5)			
			Togo	60	(0.2)			
			Tunisia	270	(0.8)			
			Peru	310	(0.9)			
			Syria	190	(0.6)			
			Venezuela	10	(..)			
			Others	300	(0.9)			
Totals	7710	(22.5)		24995	(73.0)		1540	(4.5)
Grand Total		34245						

This table uses production costs of under $100/tonne.

World resources are immense and deposits are now being discovered on the continental shelf.

PHOSPHATE

WORLD MINE PRODUCTION, 1989-90
(million tonnes and % of total 1988)

Developed	1989	1990	% 1990	Developing	1989	1990	% 1990	Centrally Planned	1989	1990	% 1990
Finland	0.58	0.55	(0.4)	Algeria	1.22	1.10	(0.8)	China	17.07	17.30	(11.2)
S Africa	2.96	3.16	(2.0)	Brazil	3.66	2.97	(1.9)	N Korea	0.50	0.50	(0.4)
Sweden	0.71	..	(...)	Egypt	1.35	1.30	(0.8)	USSR	34.40	33.50	(21.7)
USA	48.87	46.34	(30.0)	India	0.70	0.66	(0.4)	Vietnam	0.50	0.27	(0.2)
Others	0.09	0.10	(0.1)	Iraq	1.30	1.10	(0.7)				
				Israel	3.90	3.52	(2.3)				
				Jordon	6.90	5.92	(3.8)				
				Mexico	0.66	0.60	(0.4)				
				Morocco	18.07	21.40	(13.9)				
				Nauru	1.18	0.96	(0.7)				
				Senegal	2.27	2.15	(1.4)				
				Syria	2.25	1.67	(1.1)				
				Togo	3.36	2.31	(1.5)				
				Tunisia	6.61	6.26	(4.0)				
				Zimbabwe	0.13	0.16	(0.1)				
				Others	0.40	0.33	(0.2)				
Totals	53.21	50.15	(32.5)		53.96	52.41	(34.0)		52.4	51.57	(33.5)

Grand Totals 1989- 159.57
1990- 154.13

The P_2O_5 content of production was 49.4 million tonnes in 1989 (down from 50.7 million tonnes in 1988) and 47.5 million tonnes in 1990.

WORLD PRODUCTIVE CAPACITY, 1990
(million tonnes)

Developed		Developing		Centrally Planned	
S Africa	4.7	Algeria	2.3	China	19.0
USA	55	Christmas Is	1.0	N Korea	1.0
Others	2	Israel	4.0	USSR	36.0
		Jordan	8	Vietnam	1.0
		Mexico	1.0		
		Morocco & W			
		Sahara	34.1		
		Nauru	2.0		
		Senegal	2.1		
		Togo	3.2		
		Tunisia	10.0		
		Others	9.0		
Totals	61.7		76.7		57.0
Grand Total	195.4				

RESERVE/PRODUCTION RATIOS

Static Reserve Life (years): very large
Ratio of identified reserve base to
cumulative demand 1991-2010: 9 : 1

CONSUMPTION

	'000 tonnes		% p.a. growth rates	
	1989	1990	1970s	1980s
European Community	14251	13722	-0.1	-3.5
Japan	1575	1489	-0.9	-3.3
United States(a)	40188	41730	5.5	0.3
Other Western world	40326	42065	9.1	-0.8
Total Western World	**96340**	**99006**	**5.2**	**-0.9**
Total World	**159317**	**157281**	**2.3**	**1.3**

Source: Phosphorus and Potassium: British Sulphur Publishing
(a) USBM estimates consumption at 42143 in 1989 and 43967 in 1990

END USE PATTERNS, 1990 (USA)(%)

Fertilisers and animal feed supplements	95
Industrial and food grade products	5

VALUE OF ANNUAL PRODUCTION

$5.6 billion (at average of 1991 prices for Floridan and Moroccan rock).

SUBSTITUTES

No substitutes exist for agricultural applications.

The level of sodium tripolyphosphate in detergents is being reduced by substitution with other compounds.

TECHNICAL POSSIBILITIES

Mining of deep deposits.

PRICES

	1986	1987	1988	1989	1990	1991
USA Weighted average value fob mine All grades domestic and export US$/tonne	22.25	19.37	19.56	21.76	23.20	24.00
Real Dec 1989 prices	25.8	21.9	21.3	22.7	23.2	23.9
Moroccan 75-77% BPL fas Casablanca $/tonne	48.5	48.5	48.5	48.5	48.5	48.5

Note: Moroccan price is nominal only.

PHOSPHATE

Prices are fixed on a contract basis depending on quality and grade. Phosphate fertiliser contracts are usually short term in duration whilst the acid business has annual contracts with six months' pricing. Actual prices are not published and the above are only guidelines. US prices usually lag behind the Moroccan. Typically US domestic prices are much lower than those achieved in export markets, where US producers fix common prices. In 1990, for example, domestic prices averaged $21.91/tonne f.o.b. against $30.66/tonne f.o.b. for exports.

MARKETING ARRANGEMENTS

Fertiliser and acid markets are now supplied mainly by large integrated producers, normally government controlled, with captive phosphate rock. Morocco, USSR and USA account for 66% of world production, but new developments, including attendant acid and fertiliser plants, are coming onstream worldwide and are diversifying supply sources.

REAL PRICES 1979 to 1991
Phosphate Rock, Average US mine value

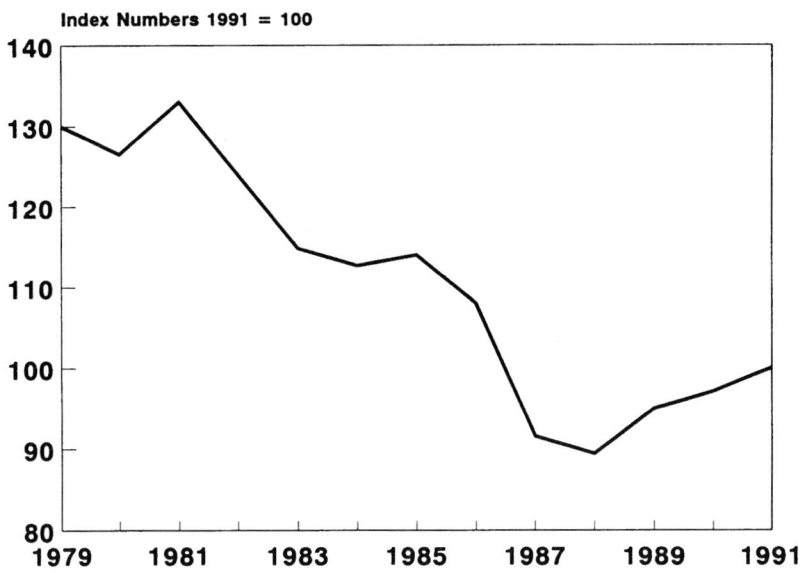

Index Numbers 1991 = 100

WORLD PRODUCTION 1979 to 1991
Phosphate Rock

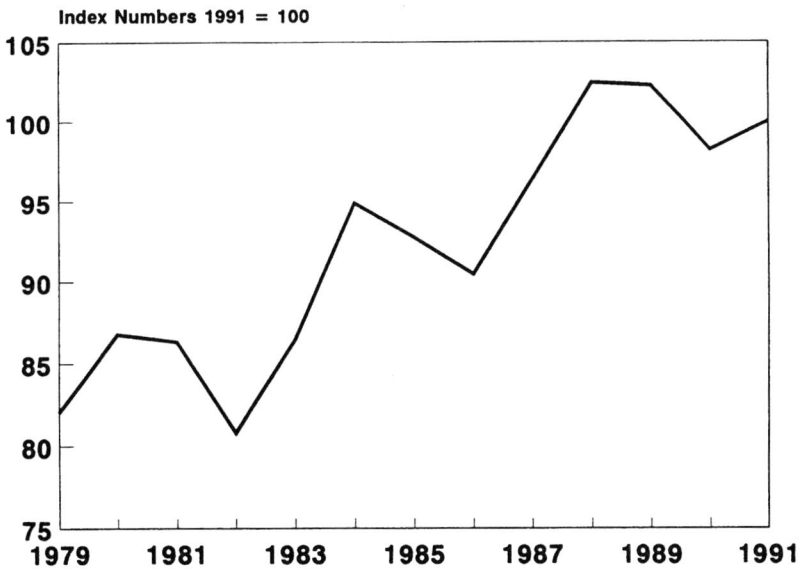

Index Numbers 1991 = 100

PHOSPHATE

SUPPLY AND DEMAND BY MAIN MARKET AREA

	UK 1989	UK 1990	EC(12) 1989	EC(12) 1990	Japan 1989	Japan 1990	USA 1989	USA 1990
Production ('000 tonnes)								
Mine production	-	-	-	-	-	-	49817	46343 (marketable)
Net Imports ('000 tonnes)								
Phosphate rock (gross)	602	548	14183	12836	1590	1543	705	451
Superphosphates	316	383	1053	1292	59	64	22	1
Basic slag	4.3	4.5	0.7	3.0	-	-	45	1
Source of Net Imports (%)								
Phosphate rock (gross)								
Canada								
European Community								1
United States	7		21	18	51	49		
Algeria			3	3				
Israel	1	1	11	11		3		
Jordan			2	2	14	15		
Morocco	79	90	40	41	14	11	99	98
Senegal	6		4	4	4	4		
S Africa		6	6	15	16			
Syria			3	4				
Togo			6	6				
Tunisia	7	8	3	2				
Others and unidentified		1	1	2	2	2	1	1
USSR				1				
Net Imports ('000 tonnes)								
Phosphate rock	0.2	0.7	21.2	10.6	-	-	7842	6238
Superphosphates	0.4	0.1	20	54	34.6	0.2	517	714
Basic slag	0.1	0.1	84	89	-	-	-	-
Consumption ('000 tonnes)								
Phosphate rock	532	531	14251	13722	1575	1489	40326 42143	42065 43967 (all forms)
Phosphate fertilisers (P_2O_5 content)	414	415	4138	3880	728	690	3942	3765

	UK		EC(12)		Japan		USA	
	1989	**1990**	**1989**	**1990**	**1989**	**1990**	**1989**	**1990**
Import Dependence								
Imports as %								
of consumption	100	100	100	100	100	100	-	-
Imports as % of consumption								
and net exports	100	100	100	100	100	100	-	-
Share of World Consumption (%)								
Phosphate Rock								
Total World	..	9	9	1	1	25	27	
Western World	1	1	15	14	2	2	42	42
Phosphate Fertilisers								
Total World	1	1	11	11	2	2	11	11
Consumption Growth (% p.a.)								
1970s	0.6		1.7(a)		0.8		4.3	
1980s	n.a	-3.5		-3.3		0.3	•	

(a) Based on rock imports.

PLATINUM GROUP

(Platinum, Palladium, Iridium, Osmium, Rhodium, Ruthenium)

WORLD RESERVES
(tonnes and % of total)

Developed			Developing		Centrally Planned		
Canada	250	(0.4)	Colombia		USSR	5900	(10.5)
S Africa	50000	(88.6)	Zimbabwe	} very small			
USA	250	(0.4)					
Others (a)	30	(0.1)					
Totals	**50530**	**(89.5)**				**5900**	**(10.5)**
Grand Total		**56430**					

(a)Includes Australia, Finland and Yugoslavia.

The different deposits of platinum group metals have markedly different ratios between the constituent metals. The US Bureau of Mines and the South African Minerals Bureau give the following breakdowns for the main deposits (in percentage by weight).

	Platinum	Palladium	Iridium	Rhodium	Ruthenium	Osmium
Colombia	93	1	3	2	-	1
Canada- Sudbury	43	45	2	4	4	2
S Africa- Merensky	61	26	1	3	8	1
- UG2	41	34	2	9	12	2
- Platreef	44	48	3	1	4	..
USSR- Norilsk	25	67	2	3	2	1
USA- Stillwater	20	78	..	1	..	-
- Duluth	18	78	1	2	1	-

The world reserve base of platinum group metals is 66,000 tonnes and world resources are around 100,000 tonnes.

WORLD MINE PRODUCTION, 1989-90, and PRODUCTIVE CAPACITY, 1990
(kilograms and % of total 1990)

	Mine Production		% of Production	Productive Capacity
	1989	1990	1990	1990
Developed				
Australia	500	500	(0.2)	600
Canada	9870	11209	(3.9)	12000
Finland	160	160	(0.1)	300
Japan (a)	1852	2419	(0.8)	-
S Africa	135800	138500	(48.3)	150000
USA	6280	7740	(2.7)	8000
Yugoslavia	162	164	(0.1)	300
Total	**154624**	**160692**	**(56.1)**	**171200**
Developing				
Colombia	964	950	(0.3)	1000
Ethiopia	2	2	(..)	
Zimbabwe	68	60	(..)	300
Total	**1034**	**1012**	**(0.3)**	**1300**
Centrally Planned				
USSR	127500	125000	(43.6)	130000
Total	**127500**	**125000**	**(43.6)**	**130000**
TOTAL	**283158**	**286704**		**302500**

(a)Japanese smelter/refinery recovery from ores originating elsewhere (including Australia, Canada, Indonesia, Papua New Guinea and Philippines), but this is not thought to result in substantial double counting. China is also believed to produce platinum group metals.

The estimated breakdown of 1990's Western World production was, in percentages:

Platinum	45
Palladium	47
Iridium	2
Ruthenium	2
Rhodium	3
Osmium	1

RESERVE/PRODUCTION RATIOS

Static Reserve Life (years):	197
Ratio of identified reserve base to cumulative demand 1991-2010:	19.5 : 1

PLATINUM GROUP

CONSUMPTION (Platinum and Palladium)

	tonnes		% p.a. growth rates	
	1989	1990	1970s	1980s
Platinum				
Western Europe	17.57	20.68	n.a	5.4
Japan	50.70	54.28	n.a	8.4
North America	30.95	29.86	n.a	-0.4
Other Western World	7.93	9.18	n.a	7.0
Total Western World	**107.15**	**114.0**	**n.a**	**5.4**
Palladium				
Western Europe	18.20	18.35	n.a.	n.a.
Japan	47.59	48.37	n.a	7.6
North America	35.0	35.46	n.a	3.8
Other Western World	5.29	6.69	n.a	6.8(inc.W.Europe)
Total Western World	**106.08**	**108.87**	**n.a**	**6.0**
Rhodium				
Western Europe	2.80	3.02	n.a	n.a
Japan	3.17	3.70	n.a	n.a
North America	3.52	3.86	n.a	n.a
Other Western World	0.90	0.84	n.a	n.a
Total Western World	**10.39**	11.42	n.a	n.a
Platinum and Palladium				
Western Europe	35.77	39.03	n.a	n.a
Japan	98.29	102.65	10.1	8.0
North America (a)	65.95	65.32	6.3	1.7
Other Western World	13.22	15.87	n.a	7.8(inc.W.Europe)
Total Western World	**213.23**	**222.87**	**n.a**	**5.7**

Source:Johnson Matthey

(a) The USA comprises 85-90% of North American consumption.

Consumption is shown gross before recycling of autocatalysts. For platinum and palladium combined, this amounted to 7.6 tonnes in 1989 and 9.3 tonnes in 1990. Demand for investment, large and small, is also excluded as it is not genuine consumption. Such 'demand' amounted to nearly 5 tonnes of platinum in 1989 and 6.2 tonnes in 1990.

Demand for ruthenium was 4.7 tonnes in 1990 and for iridium 0.8 tonnes.

END USE PATTERNS, 1990 (%)

	USA						Pt Grp.	Japan Platinum & Palladium
	Platinum	Palladium	Rhodium	Iridium	Ruthium	Osmium		
Automotive	58	9	72	..	-	-	35	14
Electrical	11	51	7	25	86	-	31	32
Chemical	6	7	4	5	1	-	6	..
Dental	2	17	..	23	..	99	9	10
Jewellery	1	1	5	2	..	-	1	37
Petroleum refining	9	4	..	-	-	-	6	-
Others	13	11	12	45	13	1	12	7

Western World %

Platinum   Jewellery 37, autocatalyst 42, investment 4, electrical 5, chemical 4, glass 4, petroleum 2, other 2.

Palladium   Electrical 49, dental 29, autocatalyst 9, jewellery 5, other 8.

Rhodium   Autocatalyst 84, chemical 7, glass 3, electrical 3, other 3.

Ruthenium   Electronics 64, electrochemical 34, other 2.

Iridium   Electrochemical 49, crucibles 10, other 41.

Source:Platinum 1992. Johnson Matthey

VALUE OF CONTAINED METAL IN ANNUAL PRODUCTION

$3.09 billion (at average 1991 prices for the various metals, weighted according to 1990 production split).

MARKETING ARRANGEMENTS

Most mining is in association with nickel-copper ores with the USSR, South Africa (Rustenburg and Impala), and to a lesser extent Canada (Inco),controlling the market. All are integrated producers. S Africa controls producer price of platinum and USSR that of palladium, and both can influence world spot price by curtailing production and purchasing excess metal. As the usage of platinum group metals in automotive exhaust catalysts matures in the USA, there will be a rising supply of secondary material. This will however be more than offset by demands for primary material in Europe as European automotive emission standards are tightened in the 1990s. Secondary autocatalyst recovery accounted for approximately 5.4% of total platinum supply in 1990, but scrap was additionally available from other uses.

SUBSTITUTES

It is usually easier to substitute metals of the platinum group for one another, especially in alloys, than to use alternate materials.

Substitutes in electrical uses include tungsten, nickel, silver, gold and silicon carbide.

PLATINUM GROUP

Alternative catalysts include nickel, molybdenum, tungsten, chromium, cobalt, vanadium, silver and rare-earth materials, but normally with efficiency and cost penalties. However rhenium has been used most satisfactorily for part of platinum in petroleum-refining catalysts.

Stainless steel and ceramics can be used where corrosion resistance is of primary concern.

TECHNICAL POSSIBILITIES

Recovery from radio-active waste and the creation of artificial platinum group metals in nuclear power reactors, but both seem improbable in the foreseeable future.

Increased recovery of PGM in automotive catalytic converters. The lean burn engine, reduced lead content in petrol and electric cars could reduce application in this field.

Use of tin-lead alloys as substitutes for precious metals in electronic applications.
A major area of growth is in fuel cells. The phosphoric acid fuel cell is technically well proven and industrial applications are developing rapidly.

PRICES

	1986	1987	1988	1989	1990	1991
$/troy oz.						
Platinum						
Nymex Spot	466.3	559.0	531.6	511.9	471.4	376.0
Real Dec 1989 prices	541.4	631.0	579.3	530.9	471.9	374.5
Palladium:NY dealer	114.9	125.5	124.7	143.7	113.9	86.8
Iridium:NY dealer	421.4	367.9	309.6	308.5	319.0	293.2
Osmium:NY dealer	753.8	633.0	588.0	548.0	436.0	425.0
Rhodium:NY dealer mean	1190.1	1243.0	1275.0	1308	3650.3	3835.7
Ruthenium:NY dealer	75.9	72.3	64.9	64.1	62.9	58.5

There is a combination of producer and dealer pricing, with futures trading in the USA and Japan. Markets are subject to speculative activity. Rhodium prices rocketed in mid-1990 to over $7000/oz because of production problems, strong demand and speculation.

REAL PRICES 1979 to 1991
Platinum, New York spot

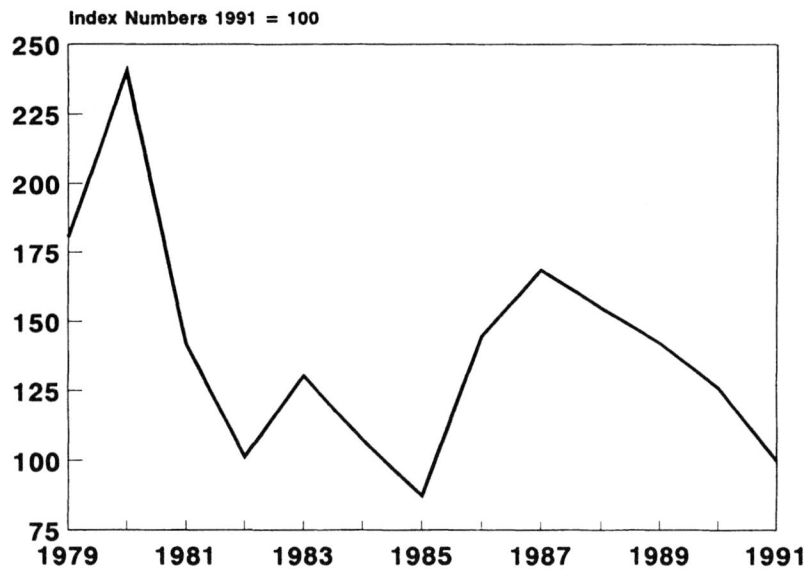

Index Numbers 1991 = 100

WORLD PRODUCTION 1979 to 1991
Platinum

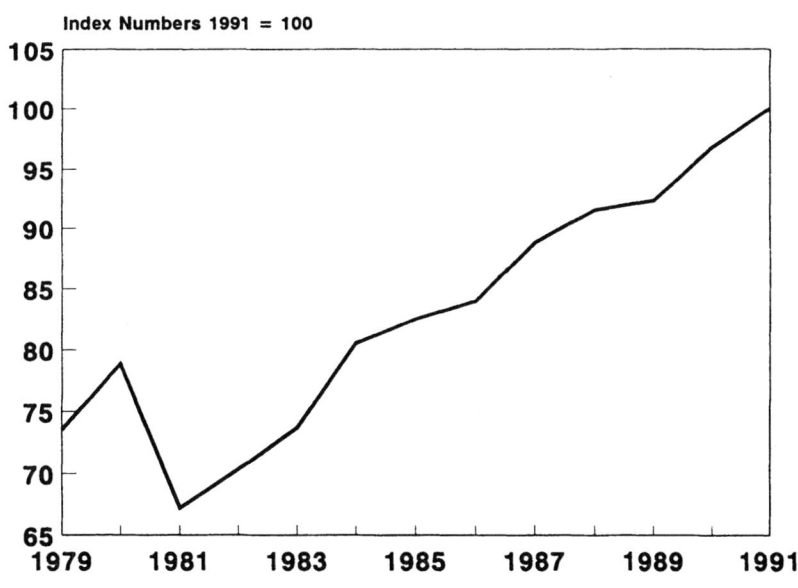

Index Numbers 1991 = 100

PLATINUM GROUP

SUPPLY AND DEMAND BY MAIN MARKET AREA

	UK 1989	UK 1990	EC(12) 1989	EC(12) 1990	Japan 1989	Japan 1990	USA 1989	USA 1990
Production (kilograms)								
Mine production	-	-	-	-	-	-	n.a	n.a
Refinery production:								
New metal	n.a	n.a	n.a	n.a	1852	c.1900	339	64
Secondary metal	n.a	n.a	n.a	n.a	-	-	50525	52311
Net Imports (kilograms of Pt group)								
Ores and concentrates	n.a	n.a	n.a	n.a	n.a	n.a	12	109
Waste and scrap	382000	310000	446000	366000	5294	5295	2162	4614
Unwrought metal, inc. alloys	46792	57009	72592	90533	58240	65527	82478	81265
Semi-manufactures, inc. alloys	4738	6520	12530	16199	4843	5167	28628	39360
Total All forms	n.a.	n.a.	n.a.	n.a.	n.a.	n.a.	113278	125354

Source of Net Imports (%)

	Unwrought/Wrought Waste & Scrap						All forms	
Australia								
Canada	5	1	5	1	2	1	2	1
EEC	12	13			32	33	26	26
Namibia	15		11					
Hong Kong								
S Africa	6		8	3	15	15	50	54
Switzerland		1	3	6	10	9	3	1
Norway	3	1	2	1				1
Sweden	2			1				
Czechoslovakia			3	3				
United States	26	63	34	64	11	5		
USSR			2	1	22	29	13	12
Mexico	31		25					
Saudi Arabia	11		9					
Zambia	4		4					
Others	2	6	5	9	8	8	6	5

Most ores and concentrates imported into the UK for refining are from South Africa, whose importance is thus much greater in world trade than this table suggests.

	UK 1989	UK 1990	EC(12) 1989	EC(12) 1990	Japan 1989	Japan 1990	USA 1989	USA 1990
Net Exports (kilograms) Unwrought & semi-manufactures inc. alloys, ores & concentrates	74139	81887	68384	69780	4504	7201	38311	55044
Consumption (kilograms)	n.a	n.a	n.a	n.a	(b) 101460	(b) 106350	92348(a)	98320(a)

(a) These are from the US Bureau of Mines and are on a different basis from the data shown earlier. They include secondary production on a non toll basis.
(b) Platinum, palladium and rhodium only.

Import Dependence

	UK 1989	UK 1990	EC(12) 1989	EC(12) 1990	Japan 1989	Japan 1990	USA 1989	USA 1990
Imports as % of consumption	100	100 (exc. secondary)	100	100	100	100	100	100
Imports as % of consumption and net exports	100	100 (exc. secondary)	100	100	100	100	100	100

Share of World Consumption (%)

Insufficient information is published to complete this section, especially bearing in mind the large secondary recovery of platinum group metals.

Consumption Growth (% p.a.)

	UK	EC(12)	Japan	USA
1970s	n.a	n.a	10.1	6.3
1980s	n.a	n.a	8.0	1.7

POTASH

WORLD RESERVES
(million tonnes K$_2$O and % of total)

Developed			Developing			Centrally Planned		
Canada	4400	(46.5)	Brazil	50	(0.5)	China	320	(3.4)
France	16	(0.2)	Chile	10	(0.1)	E Germany	300	(3.2)
W Germany	500	(5.3)	Israel	53	(0.6)	USSR	36000	(38.0)
Italy	20	(0.2)	Jordan	54	(0.6)			
Spain	28	(0.3)	Others	..	(...)			
UK	25	(0.3)						
USA	85	(0.8)						
Totals	5074	(53.6)		167	(1.8)		4220	(44.6)
Grand Total	9461							

The world's reserve base is estimated at approximately 17,200 million tonnes, of which 22% is located in the USSR. Total world resources exceed 250 billion tonnes, much of it only recoverable through solution mining techniques due to depth.

WORLD MINE PRODUCTION, 1989-90, and PRODUCTIVE CAPACITY, 1990
(million tonnes of K_2O and % of total 1990)

	Mine Production 1989	Mine Production 1990	% of Production 1990	Productive Capacity 1990
Developed				
Canada	7.07	7.37	(26.1)	11.52
France	1.20	1.30	(4.6)	1.68
W Germany	2.18	2.20	(7.8)	2.70
Italy	0.16	0.06	(0.2)	0.40
Spain	0.85	0.69	(2.4)	0.75
UK	0.48	0.51	(1.7)	0.50
USA	1.59	1.71	(6.0)	1.91
Total	**13.53**	**13.84**	**(48.9)**	**19.67**
Developing				
Brazil	0.10	0.06	0.2	0.15
Chile	0.03	0.03	0.1	0.04
Israel	1.34	1.35	4.8	1.35
Jordan	0.79	0.79	2.8	n.a
Total	**2.26**	**2.23**	**7.9**	**1.54**
Centrally Planned				
China	0.04	0.04	0.1	0.12
E Germany	3.20	2.70	9.5	2.90
USSR	10.20	9.50	33.6	13.70
Total	**13.44**	**12.24**	**43.2**	**16.72**
TOTAL	**29.2**	**28.31**		**37.33**

RESERVE/PRODUCTION RATIOS

Static Reserve Life (years):	300
Ratio of identified reserve base to cumulative demand 1991-2010:	30 : 1

CONSUMPTION
(Fertiliser uses only)

	'000 tonnes K_2O 1989	'000 tonnes K_2O 1990	% p.a. growth rates 1970s	% p.a. growth rates 1980s
European Community	4657	4459	1.5(approx.)	-0.1
Japan	568	537	1.0	0.4
United States(a)	5150	5453	4.5	-1.9
Other Countries	5773	5464	n.a	0.9
Western World	16148	15913	n.a	0.1
Total World	**26778**	**24314**	**n.a**	**1.8**

(a) Apparent Consumption all uses.
Source: Phosphorus and Potassium : British Sulphur Publishing

POTASH

END USE PATTERNS 1990 (USA)(%)

Fertiliser industry   95
Other (primarily caustic potash-
chlorine plants)   5

VALUE OF ANNUAL PRODUCTION

$2.6 billion (at average 1991 prices)

SUBSTITUTES

Potash used in industrial applications can sometimes be replaced by sodium compounds.

No substitutes for agricultural use. Manure and glauconite are low potassium content sources that can be transported short distances to the crop fields.

TECHNICAL POSSIBILITIES

Increased resort to solution mining of underground deposits.

Recovery from low grade resources or as a by-product in the production of alumina from alunite.

New industrial applications, eg: electrical plants.

Extensive and intensive cropping changes. The effect of fertiliser pollution on water supplies could reduce potash consumption capacity especially in developing countries.

PRICES

	1986	1987	1988	1989	1990	1991
Saskatchewan Standard fob bulk/short ton	49.0	56.19	83.36	84	84	83.4
Real Dec 1991 prices	56.9	63.5	91.5	87.4	84.3	83.4
Average US fob mine value $/tonne K205 muriate	82	93	132	137	130	131

Producer list pricing for long term contracts. Discounting prevalent.

In January 1988, a 5 year antidumping agreement was signed between Canadian producers and the US Department of Commerce, bringing some pro-rating of Canadian production and higher prices in the US. The agreement established a two-tier world pricing system which increased offshore imports to the US to record levels. The world price rose gradually towards the US price towards the end of the year.

MARKETING ARRANGEMENTS

USSR, E Germany, N America and W Europe have provided almost all the world's supply of potash. USSR production fluctuates considerably, with consequent effects on the world market. A high proportion of Canadian production, second after USSR, has been controlled by a provincial government (Saskatchewan). The potash market is highly dependent on the state of the world farming economy. China became a large purchaser in the late 1980s introducing a further erratic element. The main Canadian producer, PCS, is being privatised. With the unification of Germany, much East German capacity has been closed because of costs and environmental damage. Russian output is also at risk.

REAL PRICES 1979 to 1991
Potash, Average US mine value

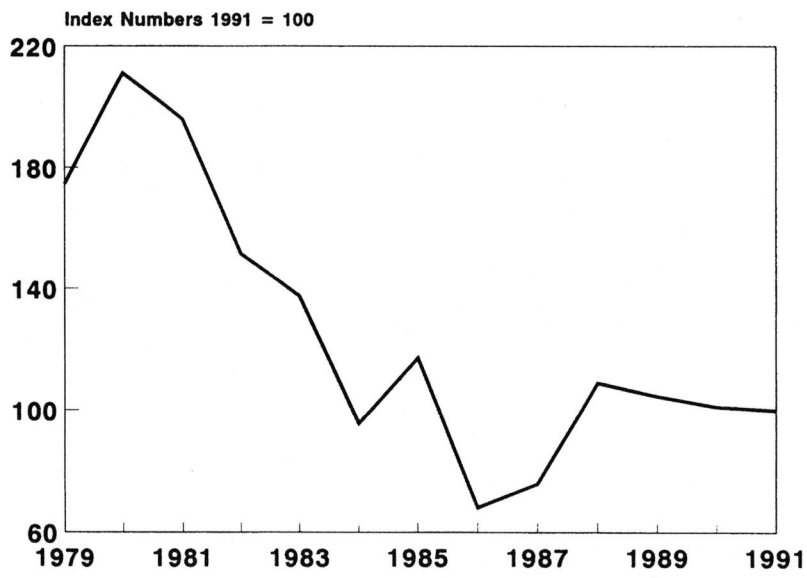

Index Numbers 1991 = 100

WORLD PRODUCTION 1979 to 1991
Potash

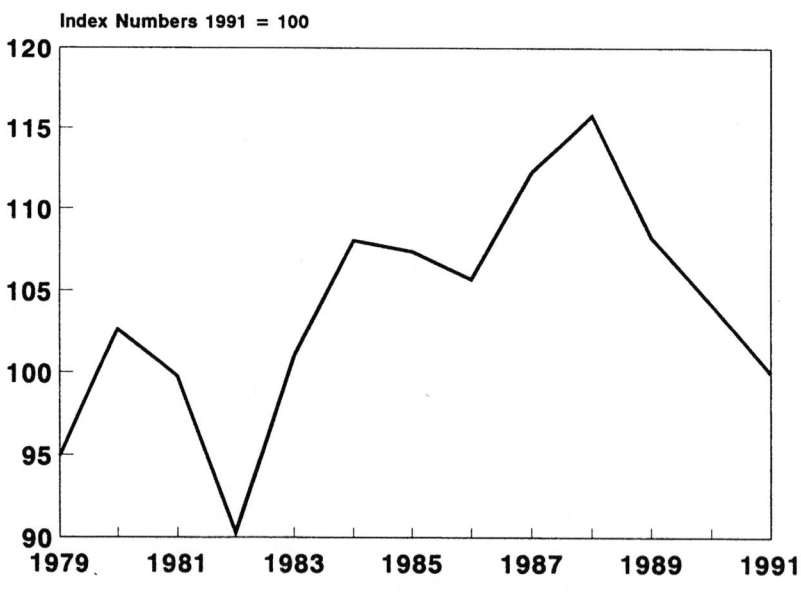

Index Numbers 1991 = 100

SUPPLY AND DEMAND BY MAIN MARKET AREA

	UK 1989	UK 1990	EC(12) 1989	EC(12) 1990	Japan 1989	Japan 1990	USA 1989	USA 1990
Production ('000 tonnes K_2O)	486	514	4869	4764	-	-	1595	1713
Net Imports ('000 tonnes)								
All forms (K_2O)	c.285	c.250	c.1650	c.1510	n.a	n.a	4050(a)	4164
Potassium chloride (gross wt)					948	944	5510	6816
Under 40% K_2O	-	0.5	2.9	0.5	n.a	n.a	n.a	n.a
40-62% K_2O	451	404	2679	2432	n.a	n.a	n.a	n.a
Over 62% K_2O	1.7	0.5	4.9	1.0	n.a	n.a	n.a	n.a
Potassium sulphate (gross wt)	11.6	9.4	20.6	46.0	306	190	63	52
Source of Net Imports (%)								
Potassium chloride & sulphate								
Canada	1	11	18	12	48	52	91	91
European Community	45	43			17	13	2	3
United States					11	7		
East Germany	45	33	18	11	3	4		
USSR	1	1	30	39	13	16	1	
Israel	7	10	28	31	1	2	5	6
Jordan			5	5	3	4		
Others	1	2	1	2	4	2	1	
Net Exports ('000 tonnes)								
All forms (K_2O)	c.225	c.235	c.430	c.1025	n.a	n.a	446	470
Potassium chloride (gross wt)					0.6	0.9	497	446
Under 40% K_2O	0.9	0.4	10.0	10.4	n.a	n.a	n.a	n.a
40-62% K_2O	365.7	382.3	950.4	1091.8	n.a	n.a	n.a	n.a
Over 62% K_2O	1.1	1.9	7.2(a)	3.0(a)	n.a	n.a	n.a	n.a
Potassium sulphate (gross wt)	0.8	0.2	521.9(b)	685.4(b)	0.7	0.1	153	244

(a) Excl. Germany
(b) Excl. Beg/Lux. & Germany

	UK 1989	UK 1990	EC(12) 1989	EC(12) 1990	Japan 1989	Japan 1990	USA 1989	USA 1990
Consumption ('000 tonnes K_2O)	515(a)	510(a)	4657	4459	568	537	5150	5453

(a) According to British Geological Survey, consumption was 560.6 in 1989 and 560.3 in 1990.

POTASH

	UK		EC(12)		Japan		USA	
	1989	**1990**	**1989**	**1990**	**1989**	**1990**	**1989**	**1990**
Import Dependence								
Imports as % of consumption	55	49	35	34	100	100	79	76
Imports as % of consumption and net exports	38	34	32	27	100	100	72	70
Share of World Consumption (%)								
Total world	2	2	17	18	2	2	19	22
Western World	3	3	29	28	4	3	32	34
Consumption Growth (% p.a.)								
1970s	-0.7		1.5 (approx)		1.0		4.5	
1980s	3.0		-0.1		0.4		-1.9	

RARE EARTH MINERALS & METALS

WORLD RESERVES

The rare earth elements are the group of 15 chemically similar elements with atomic numbers between 57 and 71 inclusive: lanthanum, cerium, praseodymium, neodymium, promethium, samarium, europium (together making up the 'light' or 'cerium' subgroup), gadolinium, terbium, dysprosium, holmium, erbium, thulium, ytterbium and lutetium (which together with yttrium, not itself a rare earth element but invariably associated with them in nature, make up the 'heavy' or 'yttrium' subgroup).

The main sources of rare earths are the ores bastnaesite (a fluorocarbonate) and monazite (a phosphate). Small quantities also occur in the mineral xenotime, but this is primarily a source of yttrium.

('000 tonnes of Rare Earth Oxides (REO) and % of total)

Developed			Developing			Centrally Planned		
Australia	5100	(6.1)	Brazil	270	(0.3)	China	43000	(51.3)
Canada	940	(1.1)	India	2300	(2.7)	USSR	450	(0.5)
USA	12600	(15.0)	Malaysia	30	(0.1)			
Others	600	(0.1)	Others	18600	(22.2)			
Totals	19240	(22.9)		21210	(25.3)		43450	(51.8)
Grand Total		83900						

The reserve base contains 93.5 million tonnes of REO.

Deposits containing rare earths exist, inter alia, in South Africa, Egypt, Thailand, Madagascar, Malawi, Malaysia, South Korea and Sri Lanka. The figures for 'others', both developed and developing, in the above table may be too low.

The rare earths occur in many other minerals and are recoverable as by-products from phosphate rock and from spent uranium leaching. World resources are thought to be very large.

The percentage rare earth contents of the two largest deposits are as follows:

	Mountain Pass USA	Baiyenebo China
Lanthanum	33.20	23.0
Cerium	49.10	50.0
Praseodymium	4.34	6.2
Neodymium	12.00	18.5
Samarium	0.79	0.8
Europium	0.12	0.2
Gadolinium	0.17	0.7
Others	Balance	Balance

RARE EARTH MINERALS & METALS

WORLD MINE PRODUCTION, 1989-90, and PRODUCTIVE CAPACITY, 1990
(tonnes of REO content and % of total 1990)

	Mine Production 1989	Mine Production 1990	% of Production 1990	Productive Capacity 1990
Developed				
Australia	7400	7975	(14.9)	11020
Canada	100	...	(...)	1000
USA (inc monazite)	21237	23163	(43.2)	26500
Total	**27737**	**31138**	**(58.1)**	**38520**
Developing				
Brazil	1900	1100	(2.0)	2200
India	2200	2475	(4.6)	2200
Malaysia	1646	1925	(3.6)	3900
Sri Lanka	110	110	(0.2)	110
Thailand	365	358	(0.7)	1175
Zaire	96	94	(0.2)	55
Total	**6317**	**6062**	**(11.3)**	**9640**
Centrally Planned				
China	25220	16480	(30.6)	31050
USSR	n.a.	n.a.	(...)	1500
Total	**25220**	**16480**	**(30.6)**	**32550**
TOTALS	**60274**	**53680**		**80710**

Indonesia, Mozambique, S. Africa and N Korea may also produce rare earth concentrates.

RESERVE/PRODUCTION RATIO

The data are not sufficiently complete for precise estimates. The varying proportions of the different elements in each deposit also reduce the relevance of overall averages. These are, nonetheless, roughly as follows:

Static Reserve Base Life (years):	very substantial
Ratio of reserve base to cumulative demand 1991-2010:	about 50 : 1

CONSUMPTION

	(tonnes of REO) 1989	(tonnes of REO) 1990	% p.a. growth rates 1970s	% p.a. growth rates 1980s
European Community	n.a	n.a	n.a	n.a
Japan	5276	5442	n.a	n.a
USA (apparent)	27700	28741	5.2	4.8

END USE PATTERNS, 1990 (USA)(%)

Petroleum cracking catalysts	53
Metallurgical uses	22
Ceramics and glass	18
Others, including phosphors, electronics, nuclear energy and lighting	7

SUBSTITUTES

Available in many applications but usually at the expense of performance.

VALUE OF CONTAINED METAL IN ANNUAL PRODUCTION

The wide variety of ores and products makes any overall value rather misleading. On the basis of prices of concentrate (bastnaesite and monazite), however, the value of mined concentrates was $105 million at average 1991 prices.

TECHNICAL POSSIBILITIES

Increased use in X-ray screens, glass screens of colour TV tubes, fluorescent lamps, permanent magnets and electronics (including computers).

Use in alloys to store hydrogen in fuel cells and heat exchangers and as cryogenic refrigerants. Potential uses in superconducting materials.

PRICES

	1986	1987	1988	1989	1990	1991
Bastnaesite concentrates, leached 70% REO US c//lb	105	105	105	105	105	105
Monazite fob Australia min 55% REO US$/tonne	577.6	514.5	565.0	640.4	651.9	536.7
Real Dec 1991 prices	670.6	581.4	616.4	666.2	653.9	534.5

Most rare earths are sold in the form of mixed rare earth compounds. Prices are usually set by major producers but vary widely according to purity, source, availability, size of order and nature of contract.

RARE EARTH MINERALS & METALS

Rhone Poulenc's end 1990 prices in $/kg for rare earth oxides were:

Cerium	28.5
Dysprosium	132
Erbium	190
Europium	1650
Gadolinium	136.5
Holmium	510
Lanthanum	23
Lutetium	7000
Neodymium	19.7
Praseodynium	38.85
Samarium	175
Terbium	880
Thulium	3600
Ytterbium	230
Yttrium	100.5

MARKETING ARRANGEMENTS

Principal producers of rare earth concentrates are the US, Australia, Malaysia, Brazil, India, China and USSR, accounting for 95% of world REO production. Refinery production and consumption are concentrated in the US, UK, Japan, France (and possibly, USSR). High purity compounds and metal are traded largely among the industrialised countries.

REAL PRICES 1979 to 1991
Rare Earths, Monazite fob Australia

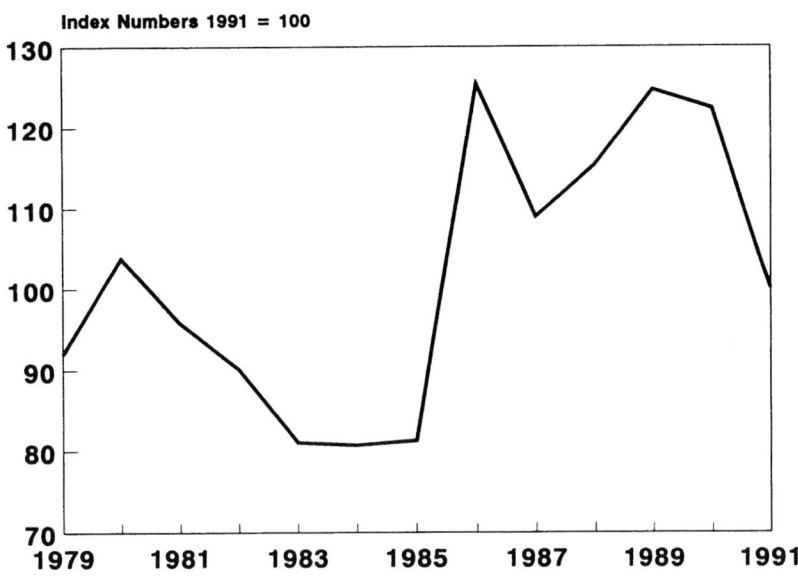

WORLD PRODUCTION 1979 to 1991
Rare Earth Concentrates

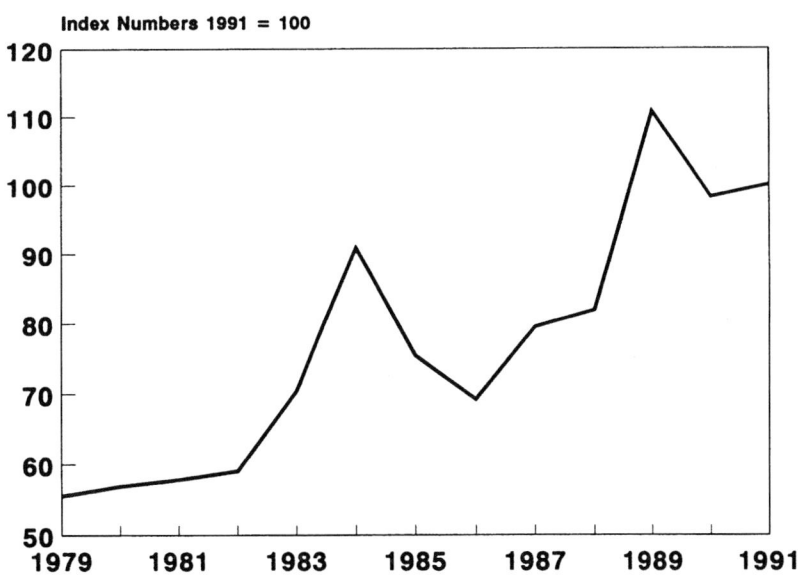

RARE EARTH MINERALS & METALS

SUPPLY AND DEMAND BY MAIN MARKET AREA

	UK 1989	UK 1990	EC(12) 1989	EC(12) 1990	Japan 1989	Japan 1990	USA 1989	USA 1990
Mine Production (tonnes REO)	-	-	-	-	-	-	21237(a)	23163(a)

(a) Mainly bastnaesite but includes an estimate of 450 tonnes/year monazite.

Net Imports (tonnes)								
Monazite	-	-	15793	16240	-	-	774	800
Rare earth oxides	n.a	n.a	n.a	n.a	1738	1557	501	154
Rare earth metals inc. alloys	77	22	573	1035	337	180	294	182
Ferrocerium & other pyrophoric alloys	385	23	122	156	498	383	85	94
Other rare earth compounds	524	592	7167	5342	9974	7337	9044	6443

Source of Net Imports (%)

Monazite								
India				7				
Australia			79	75			23	
Indonesia							77	100
South Africa			10				47	10
USA			5	4				
Others			6	14				

Rare earth oxides								
Taiwan								3
European Community					24	23	35	30
Japan							2	17
Norway								1
USA					22	19		
India							58	
China					46	35		44
USSR					8	22	1	3
Others						1	4	2

Metals and Alloys								
Austria	32		28		1	2		11
Canada						7	2	
European Community	6	27			2	1	7	6
USA		14	5	6	5	11		
Brazil	6		22	11	21	8	47	29
China	25		3		72	71	37	50
USSR				61			5	1
Japan							2	2
Others and undefined	31	59	42	22			5	1

212

	UK 1989	UK 1990	EC(12) 1989	EC(12) 1990	Japan 1989	Japan 1990	USA 1989	USA 1990
Net Exports								
(tonnes)								
Ores and concentrates	-	-	-	-	n.a	n.a	n.a	n.a
Metals and alloys	100	84	78	68	2.4(a)	11.4	425	201
Ferrocerium and other								
pyrophoric alloys	54	10	237	266	110	221	3496	2033
Other rare earth compounds	677	781	324(b)	360(b)	1069	801	4695	3815

(a)Including alkali and alkaline earth metals.
(b)Excluding France, the major European producer.

	UK 1989	UK 1990	EC(12) 1989	EC(12) 1990	Japan 1989	Japan 1990	USA 1989	USA 1990
Consumption								
(tonnes) (REO content)	n.a	n.a	n.a	n.a	5276	5442	27700	28741
Import Dependence								
Imports as % of consumption	100	100	100	100	100	100	-	-
Imports as % of consumption								
and net exports	100	100	100	100	100	100	-	-
Share of World Consumption (%)								
Total world	n.a	n.a	n.a	n.a	n.a	n.a	46	53
Consumption Growth (% p.a.)								
1970s	n.a		n.a		n.a		5.2	
1980s	n.a		n.a		n.a		4.8	

RHENIUM

WORLD RESERVES
(tonnes of metal and % of total)

Developed			Developing			Centrally Planned		
Canada	32	(1.3)	Chile	1306	(53.3)	USSR	594	(24.2
USA	386	(15.7)	Peru	45	(1.8)			
			Others	91	(3.7)			
Totals	418	(17.0)		1442	(58.8)		594	(24.2
Grand Total		2735						

Rhenium is obtained as a by-product of molybdenite in porphyry copper operations. The reserve base is 10,300 tonnes, of which over 80% is in the USA, Chile and Canada. Only a few copper/molybdenum deposits contain economically viable traces of rhenium.

WORLD MINE PRODUCTION, 1989-90
(kilograms and % of total 1990)
Note:These figures refer to recoverable rather than contained production.

Developed	1989	1990	% 1990	Developing	1989	1990	% 1990	Centrally Planned	1989	1990	% 1990
Canada	3610	3430	(12)	Chile	5000	5000	(18)	USSR	6800	6800	(24)
USA(a)	9100	11220	(41)	Peru	2000	900	(3)				
				Others(b)	550	500	(2)				
Totals	12710	14650	(53)		75001	64005	(23)		6800	6800	(24)
Grand Totals	1989-	27010									
	1990-	27850									

(a) USBM gives production of contained rhenium at 17500 tonnes/year but indicates that recoverable quantities were considerably less.
(b) Iran and Mexico.

REFINERY CAPACITY BY FINAL PRODUCER, 1990
(kilograms of contained rhenium)

	Refinery Capacity(a)
Belgium	140
Chile	9525
Finland	410
France	140
W Germany	3630
Japan	1150
Sweden	2265
United Kingdom	225
United States	4750-6580
Total Western World	**22235-24065**
Poland	100
USSR	6800
Total World	**29135-30965**

(a) Some plants may be on standby and effective capacity depends on the rhenium content of ores roasted.

RESERVE PRODUCTION RATIOS

Static Reserve Life (years)	88
Ratio of identified reserve base to cumulative demand 1991-2010	12.4 : 1

CONSUMPTION

Data on rhenium consumption are scarce. United States consumption was about 8200 kilograms in 1989 and 7700 kilograms in 1990. It increased at an average compound rate of 7.2% per annum in the 1970s, and 8.8% p.a. in the 1980s. Japanese consumption is around 1250-1500 kg/year and Western European 1750-2000 kg/year.

END USE PATTERNS 1990 (USA)(%)

Petroleum refining	45
High temperature superalloys	45
Others	10

VALUE OF CONTAINED METAL IN ANNUAL REFINED PRODUCTION

$39 million (contained metal at average 1991 prices).

RHENIUM

SUBSTITUTES

Non-rhenium catalysts are becoming more common. Iridium, gallium, germanium and silicon are among the metals being evaluated.

Substitutes in other applications are cobalt and tungsten for coatings on X-ray tubes, rhodium and rhodium-iridium for high temperature thermocouples, tungsten and platinum-ruthenium for coatings on electrical contacts and tungsten and tantalum for electron emitters.

TECHNICAL POSSIBILITIES

Use in high temperature applications such as nickel-base alloys, especially in aircraft engine components, is expanding rapidly.

Radiation screens, semiconductors, resistors, small electromagnets, heat shields, diverse catalytic reactions are all possible new uses.

Changes in petroleum refining techniques are having a detrimental effect on rhenium consumption at refineries but there is potential for a significant increase in the use of rhenium in car exhaust catalysts with the move to unleaded petrol. So far platinum group catalysts appear to dominate this market.

PRICES

	1986	1987	1988	1989	1990	1991
US Metal Powder 99.99% $/lb	350	500	680	680	680	635
US Metal Powder Real Dec 1991 prices	406.4	565	742	708	681	632
Ammonium perrhenate $/lb	200	200	500	635	635	544

Rhenium is largely a by-product of molybdenite which itself is recovered with or from porphyry copper ores. Production therefore is mainly dependent on the Cu-Mo industry. Demand is heavily dependent on the requirements of the petroleum industry. Dealer market.

MARKETING ARRANGEMENTS

Main sources of ore are Chile, Canada, USA and USSR, but recovery is concentrated in USA, Germany, Sweden, Chile and USSR. Refined output is in the hands of about ten producers, some of whom buy in concentrates for roasting.

REAL PRICES 1979 to 1991
Rhenium, US metal powder

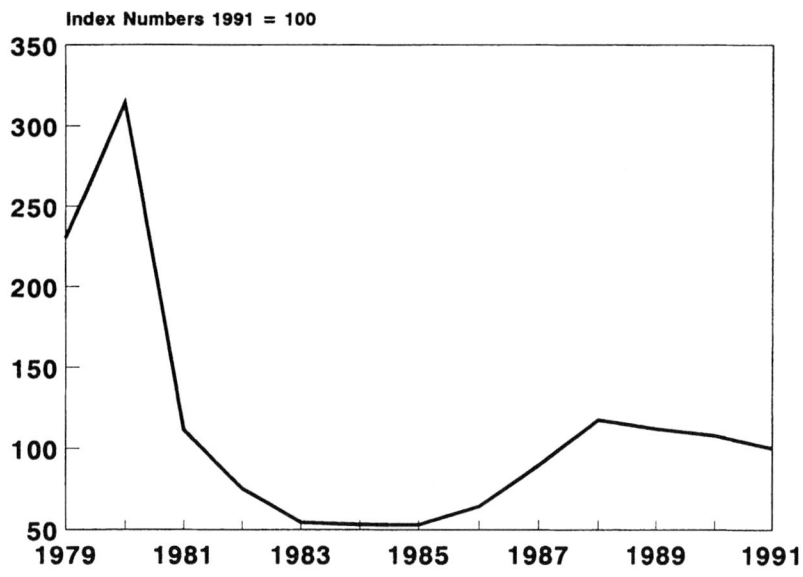

WORLD PRODUCTION 1979 to 1991
Rhenium

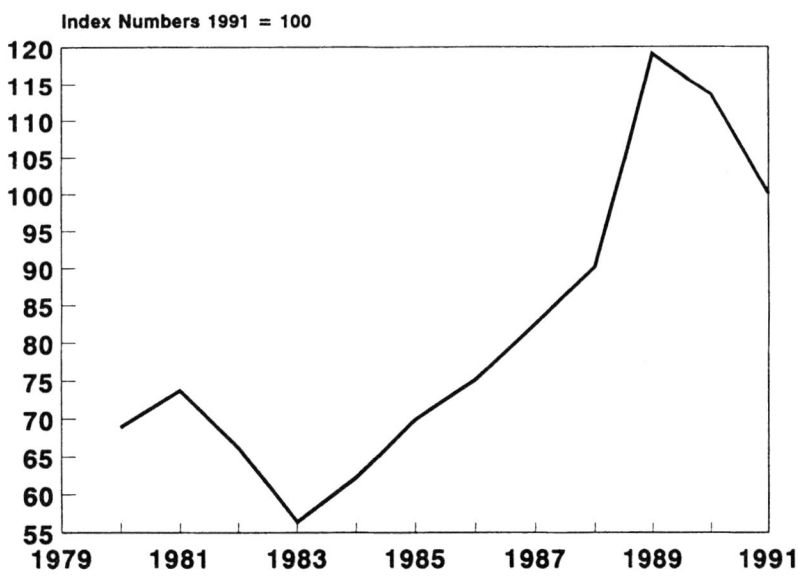

RHENIUM

SUPPLY AND DEMAND BY MAIN MARKET AREA

	UK 1989	UK 1990	EC(12) 1989	EC(12) 1990	Japan 1989	Japan 1990	USA 1989	USA 1990
Production (kilograms)								
Mine output	-	-	-	-	-	-	9100	11220
Refined output	n.a	n.a	n.a	n.a	n.a	n.a	n.a	n.a
Net Imports (kilograms of contained rhenium)								
Metal	23026(a)	18449(a)	n.a	inc. with niobium	n.a	n.a	3780	5888
Ammonium perrhenate	n.a	n.a	n.a	n.a	n.a	n.a	32417	15497

(a)Includes niobium

	UK 1989	UK 1990	EC(12) 1989	EC(12) 1990	Japan 1989	Japan 1990	USA 1989	USA 1990
Source of Net Imports (%) (metal, perrhenic acid & ammonium perrhenate)								
European Community	Most			Figures not available (ores mainly from Canada, Chile & Peru)		Figures not available	17	37
Sweden							22	8
Chile							54	18
Japan							4	3
Others							3	34
Net Exports (kilograms)								
Metal		113(a)	2600(a)	n.a	n.a	incl. in gallium	n.a	n.a

(a)Includes niobium

	UK 1989	UK 1990	EC(12) 1989	EC(12) 1990	Japan 1989	Japan 1990	USA 1989	USA 1990
Consumption (kilograms contained rhenium)	n.a	n.a	2000	c.1750-2000	c.1750-1500	c.1250-1500	c.1250-8200	7700
Import Dependence								
Imports as % of consumption	100	100	100	100	100	100	-	-
Imports as % of consumption and net exports	100	100	100	100	100	100	-	-
Share of World Consumption (%)								
Western world	n.a	n.a	c.6-7	c.6-7	c.5-6	c.5	c.30	c.28
Consumption Growth (% p.a.)								
1970s	n.a		n.a		n.a		7.2	
1980s	n.a		n.a		n.a		8.8	

SELENIUM

WORLD RESERVES
('000 tonnes of metal and % of total)

Developed			Developing			Centrally Planned		
Australia	2	(2.5)	Chile	18	(22.5)	Poland	2	(2.5)
Canada	8	(10.0)	India	1	(0.8)	USSR	4	(5.0)
USA	12	(15.0)	Mexico	4	(5.0)	Others	2	(2.5)
Others	3	(3.7)	Papua N. Guinea	1	(0.8)			
			Peru	2	(2.5)			
			Philippines	2	(2.5)			
			Zaire	8	(10.0)			
			Zambia	3	(3.7)			
			Others	8	(10.0)			
Totals	25	(31.2)		47	(58.8)		8	(10.0)
Grand Total		80						

The reserve base is 130,000 tonnes. Selenium occurs as a by-product with copper, and the above figures only cover the estimated content of economic copper deposits. Substantial resources exist in association with other metals and coal deposits and in currently uneconomic copper deposits.

WORLD REFINERY PRODUCTION, 1989-90
(tonnes of metal and % of total 1990)

Selenium is recovered mainly from the anode slimes obtained from electrolytic refining of copper. Because the selenium content of copper ore varies widely it is impossible to estimate mine production accurately. The following figures cover refinery output.

Developed	1989	1990	% 1990	Developing	1989	1990	% 1990	Centrally Planned	1989	1990	% 1990
Australia	7	7	(0.4)	Chile	47	50	(2.6)	China	100	100	(5.1)
Belgium	250	250	(12.8)	India	4	4	(0.2)	USSR	100	100	(5.1)
Canada	213	389	(19.9)	Mexico	20	12	(0.6)				
Finland	20	31	(1.6)	Peru	5	5	(0.3)				
W Germany	100	110	(5.6)	Zambia	21	22	(1.1)				
Japan	470	495	(25.4)								
Sweden	30	30	(1.5)								
USA	253	287	(14.7)								
Yugoslavia	60	60	(3.1)								
Totals	1403	1659	(85.0)		97	93	(4.8)		200	200	(10.2)
Grand Totals	1989	1700									
	1990	1952									

Selenium is recovered in Canada and the UK from used electronic and photocopier components and recycled. The USA exports material to UK for recovery.

SELENIUM

WORLD REFINERY CAPACITY, 1990

Western World capacity is split as follows:

Developed		Developing	
Japan	540	Philippines	70
USA	400	Mexico	45
Canada	450	Chile	40
Belgium	360	Zambia	25
Finland	90	Brazil	20
Sweden	70	Peru	20
Yugoslavia	50		
	1960		**220**
Total	**2180**		

Some refineries, eg in Belgium, Canada and Japan, recover selenium from secondary sources as well as from copper ores and concentrates.

RESERVE/PRODUCTION RATIOS

Static Reserve Life (years):	41
Ratio of identified reserve base to cumulative demand 1991-2010:	3.9 : 1

CONSUMPTION

	tonnes		% p.a. growth rates	
	1989	1990	1970s	1980s
Europe	830	800	n.a	c.2.5-3
Japan	283	300	1.0	3.4
United States	560	530	-5.0	3.8
Other Countries	500	500	n.a	n.a
Total World	**2173**	**2130**	n.a	n.a

Source: Mining Annual Review 1991

END USE PATTERNS, 1990 (%)
USA

		W. World	
Electronic and photocopier components	35	Electronics	25
Glass manufacturing	30	Glass	40
Chemicals and pigments	20	Pigments	10
Other, including metallurgy and agriculture	15	Metallurgy	10
		Agricultural/Biological	5
		Others	10

VALUE OF CONTAINED METAL IN ANNUAL PRODUCTION

$43 million (refined metal at 1991 average prices).

SUBSTITUTES

Substitutes exist in most end uses. Organic chemicals are used in photocopying machines, silicon substitutes in rectifier applications and cerium in glass manufacturing.

TECHNICAL POSSIBILITIES

Hydro-metallurgical processes for leaching copper sulphide concentrates to recover copper that would not allow selenium recovery.

Increased recovery from flue dust and scrap. New uses utilising electrophotographic properties. Toxicity will limit use in pigments.

PRICES

	1986	1987	1988	1989	1990	1991
European free market cif US$/lb	5.53	5.77	9.73	6.59	5.42	5.10
Real Dec 1991 prices	6.42	6.52	10.62	6.86	5.43	5.08

Source: Metal Bulletin

Selenium is derived from anode slimes obtained from refining of copper and production is therefore independent of demand. Both producer pricing and a dealer market have coexisted, with dealer prices gaining in relative importance.

MARKETING ARRANGEMENTS

Canada, Chile and the USA are the largest mine producers, Japan, Belgium, Canada and the USA the largest refinery producers. The Selenium-Tellurium Development Association promotes interest in new uses of these two metals.

REAL PRICES 1979 to 1991
Selenium, European Free Market

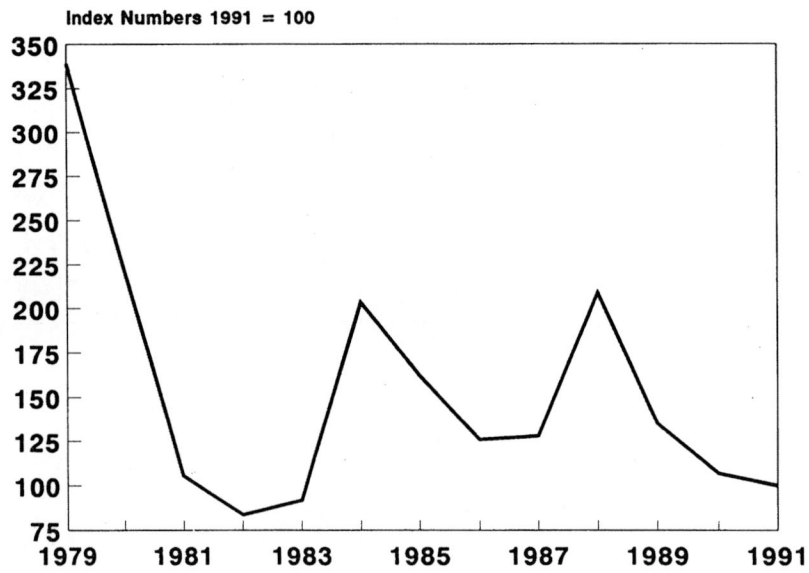

WORLD PRODUCTION 1979 to 1991
Selenium

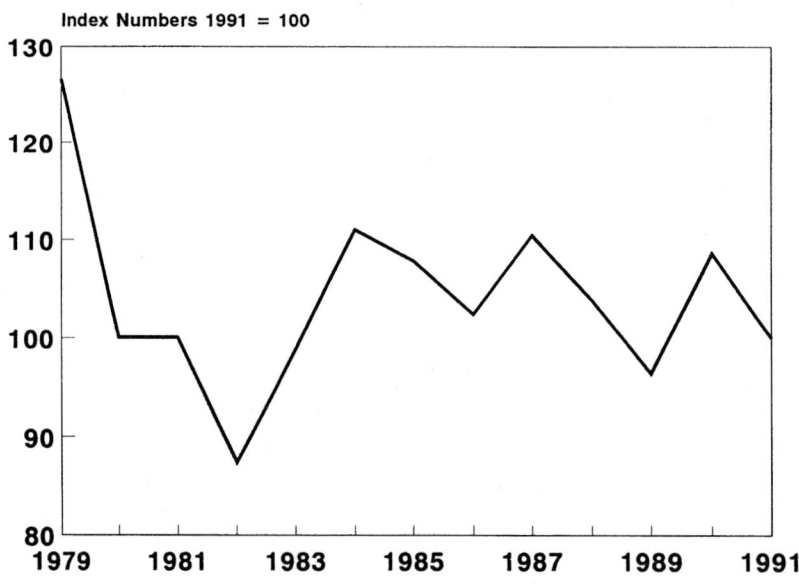

SUPPLY AND DEMAND BY MAIN MARKET AREA

	UK		EC(12)		Japan		USA	
	1989	1990	1989	1990	1989	1990	1989	1990
Production (tonnes contained selenium)								
Mine production	-	-	-	-	-	-	n.a	n.a
Refinery production Primary	-	-	c.350	c.360	470	495	253	287
Secondary	n.a.	n.a	n.a	n.a	64	60	-	-
Net Imports (tonnes contained selenium)								
Metal	637	658	657	679	58	44	400	378
Source of Net Imports (%)								
Metal								
Canada	22	19	30	25			26	30
European Community	29	12			16	2	40	41
Japan	9	15	13	17			16	14
United States		27	31	40	30		..	
Yugoslavia							10	5
Philippines					78	61	3	8
South Korea					6	35	3	2
Others and unidentified	13	23	17	28		2	2	..
Net Exports (tonnes contained selenium)								
Metal	226	294	124(a)	142(a)	374	368	372	461

(a) Excl. Belgium and Germany. USA imported 84 from Belgium and Germany in 1989.

	UK		EC(12)		Japan		USA	
Consumption (tonnes contained selenium)	c.400	c.360	c.750	c.720	283	300	560	530
Import Dependence								
Imports as % of consumption	100	100	88	94	16	15	71	71
Imports as % of consumption and net exports	100	100	68	70	9	7	43	38

If allowance is made for the import of the original raw material both the European Community and Japan rely almost wholly on imports.

Share of World Consumption (%)								
Total world	18	17	35	34	13	14	26	25

Consumption Growth (% p.a)

	UK	EC(12)	Japan	USA
1970s	2.6 (based on net imports)	n.a	1.0	-5
1980s	c.5.7 (based on net imports)	c.2.5-3	3.4	3.8

SILICON

WORLD RESERVES

Silicon is an important constituent of quartzite and other sandstones. There are ample reserves in most major producing countries in relation to demand. Estimates of total reserves, and of their geographical distribution, are not available.

WORLD PRODUCTION OF SILICON METAL, 1989-90
('000 tonnes and % of total 1990)

Developed	1989	1990	% 1990	Developing	1989	1990	% 1990	Centrally Planned	1989	1990	% 1990
Australia	9	33	(3.9)	Argentina	10	10	(1.1)	China	100	130	(14.9)
Canada	106	100	(11.5)	Brazil	117	131	(15.1)	Czecho-			
France	72	70	(8.0)	India	1	1	(0.1)	slovakia	5	5	(0.6)
Italy	19	18	(2.1)					E Germany	3	3	(0.3)
Japan	2	2	(0.2)					Hungary	2	2	(0.2)
Norway	109	77	(8.9)					Poland	10	10	(1.1)
S.Africa	36	36	(4.1)					USSR	65	65	(7.5)
Spain	11	9	(1.0)					Romania	5	5	(0.6)
Switzerland	2	2	(0.2)								
USA	136	141	(16.2)								
Yugoslavia	16	20	(2.3)								
Totals	**518**	**508**	**(58.4)**		**128**	**142**	**(16.3)**		**190**	**220**	**(25.3)**
Grand Totals	1989-	836									
	1990-	870									

WORLD PRODUCTIVE CAPACITIES, SILICON METAL AND FERROSILICON, 1990
('000 tonnes of silicon content)

Developed		Developing		Centrally Planned	
Australia	58	Argentina	37	China	830
Canada	94	Brazil	407	N Korea	25
France	152	Egypt	45	USSR	980
W Germany	60	India	60	E Europe	197
Iceland	50	Mexico	7		
Italy	95	Philippines	18		
Japan	82	S Korea	18		
Norway	418	Taiwan	10		
Portugal	12	Venezuela	48		
S Africa	105	Zimbabwe	5		
Spain	69	Others	17		
Sweden	38				
USA	468				
Yugoslavia	135				
Others	5				
Totals	**1841**		**682**		**2032**
Grand Total	**4555**				

This table covers the capacity to produce silicon metal and silicon-containing ferroalloys. These include not only ferrosilicon but also silvery pig iron and other silicon additives. Any split between the different products can only be approximate, but silicon metal producing capacity is of the order of 1 million tonnes. Some capacity, especially in the Western Countries, may be on standby.

WORLD PRODUCTION OF FERROSILICON, 1989-90
('000 tonnes and % of total 1990)

Developed	1989	1990	% 1990	Developing	1989	1990	% 1990	Centrally Planned	1989	1990	% 1990
Australia	20	20	(0.4)	Argentina	28	18	(0.4)	Bulgaria	14	14	(0.3)
Canada	88	90	(1.8)	Brazil	287	229	(4.5)	China	800	800	(15.9)
France	190	180	(3.6)	Chile	6	6	(0.1)	Czecho-			
W Germany	130	110	(2.2)	Egypt	8	8	(0.2)	slovakia	21	20	(0.4)
Iceland	72	63	(1.2)	India	74	75	(1.5)	E Germany	25	24	(0.5)
Italy	65	65	(1.3)	Mexico	9	25	(0.5)	Hungary	9	9	(0.2)
Japan	75	63	(1.2)	Philippines	9	10	(0.2)	N Korea	30	30	(0.6)
Norway	401	397	(7.9))	S Korea	5	2	(0.1)	Poland	83	89	(1.8)
S.Africa	119	107	(2.1)	Taiwan	19	16	(0.3)	USSR	1873	1860	(36.9)
Spain	38	37	(0.7)	Venezuela	55	55	(1.1)	Romania	50	50	(1.0)
Sweden	19	20	(0.4)	Peru	1	1	(...)				
Switzerland	3	3	(0.1)	Uruguay	..	..	(...)				
Turkey	5	5	(0.1)								
USA	475	434	(8.6)								
Yugoslavia	122	110	(2.2)								
Totals	**1822**	**1704**	**(33.8)**		**501**	**445**	**(8.8)**		**2905**	**2896**	**(57.4)**
Grand Totals	**1989-**	**5228**									
	1990-	**5045**									

SILICON

The silicon content of ferrosilicon varies widely from 25% to over 80%. The two most common standard grades have around 48% and 76% contained silicon. Capacities in the previous table show estimated silicon contents.

RESERVE/PRODUCTION RATIOS

For practical purposes so large as to be infinite.

CONSUMPTION

	'000 tonnes		% p.a. growth rates	
	1989	1990	1970s	1980s
Silicon metal				
European Community				
(apparent)	228	229	n.a	2.6
Japan	135	134	10.3	5.9
United States(reported)	181	198	6.9	5.4
Ferrosilicon (gross weight)				
European Community				
(apparent)	662	756	c.2	-0.1
Japan	479	535	5	2.2
United States(reported)	327	447	0.8	-0.1

END USE PATTERNS 1990, (USA) (%)

Transport	33
Machinery	14
Construction	12
Chemicals	26
Other	15

This covers the usage of silicon in all forms. Most silicon metal is used in the aluminium and chemical industries, and ferrosilica in steel and ferrous foundries.

VALUE OF ANNUAL PRODUCTION

$5.5 billion (at average 1991 prices of silicon and ferrosilicon).

SUBSTITUTES

Aluminium is among the alternatives for ferrosilicon as a deoxidiser in steel but at higher cost and production of side effects. Aluminium-silicon alloys can be replaced by some other aluminium alloys.

Germanium can be used in semiconductor and infra-red applications.

TECHNICAL POSSIBILITIES

Expansion of use in alloys particularly as a substitute for expensive additives such as chromium. Research in electronics is increasing demand for high purity silicon. Development of economically competitive silicon photovoltaic cells would increase demand also, although no major breakthrough seems likely at the moment.

Silicon faces a potential threat in the mass electronic chip market from gallium arsenide, but this is a small market for silicon. The total demand for high grade silicon from which chips are made is some 5 to 7000 tonnes/year in the Western world.

Further development of high performance silicon-based ceramics as substitutes for superalloys and other metals in high temperature or highly corrosive situations.

PRICES

	1986	1987	1988	1989	1990	1991
Metal						
UK 98% min						
£/tonne	861.4	865	865	851.4	768.8	655.0
(Source: Metal Bulletin)						
US 0.35% Fe/0.07% Ca						
cents/lb	68.1	68.1	70.4	71.5	66.3	67.0
Real Dec 1991 prices	79.1	77.0	76.8	74.4	66.5	66.8
Ferrosilicon						
US Dealer						
75-77% Si cents/lb	33.7	37.0	56.8	48.8	39.9	37.0
Real Dec 1991 prices	39.1	41.8	62.0	50.9	40.0	36.8

Prices are mainly determined on a contract basis of 3-6 months. Energy costs are important.

MARKETING ARRANGEMENTS

A wide range of companies is involved from integrated producers to one phase operators. Increasingly, ferrosilicon and silicon metal smelters are being located in low power cost countries. There is also a tendency towards plant specialisation.

Capacity was greatly increased in the 1990s. In addition the USSR began to export large tonnages. Chinese and Russian exports disrupted the European and Japanese markets in 1989-91, enforcing closures of high cost plants.

The European Community signed a minimum price agreement with the USSR in early 1991. Antidumping actions have also been successfully pursued in Europe and Japan.

REAL PRICES 1979 to 1991
Silicon, US metal

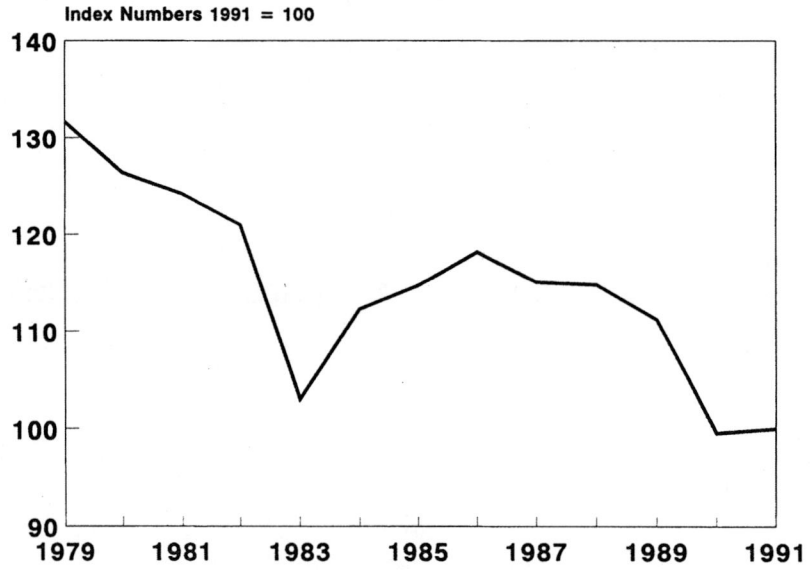

Index Numbers 1991 = 100

WORLD PRODUCTION 1979 to 1991
Silicon, All Forms

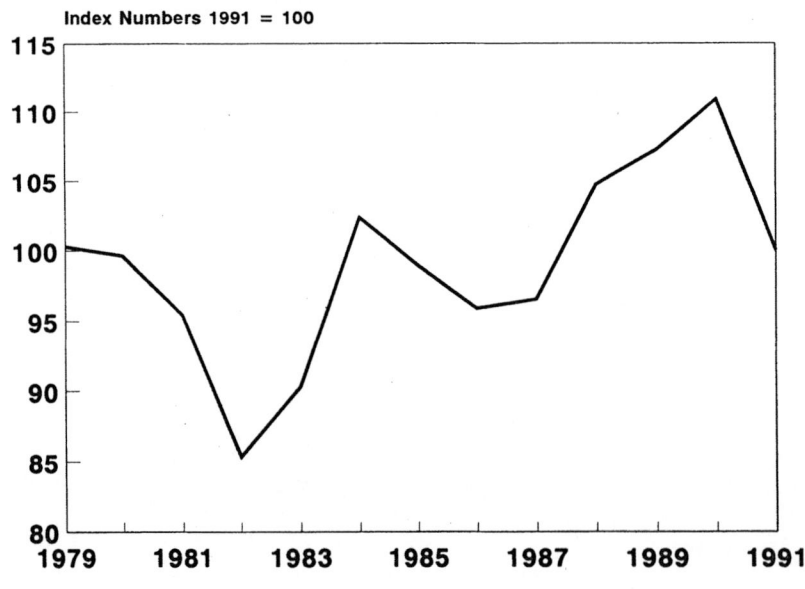

Index Numbers 1991 = 100

SUPPLY AND DEMAND BY MAIN MARKET AREA

	UK 1989	UK 1990	EC(12) 1989	EC(12) 1990	Japan 1989	Japan 1990	USA 1989	USA 1990
Production ('000 tonnes)								
Silicon metal	-	-	102	97	-	-	136	141
Ferrosilicon	-	-	423	392	75	63	475	434
Net Imports ('000 tonnes)								
Silicon metal	42.1	44.2	132.2(a)	135.5(a)	135.4	131.3	43.2	66.4
Ferrosilicon	88.1	97.7	265.3	380.4	406.9	476.5	177.9	239.0
							114.4	151.6
							(Si content)	

(a) Excluding Denmark

Source of Net Imports (%)

Silicon Metal	UK 1989	UK 1990	EC(12) 1989	EC(12) 1990	Japan 1989	Japan 1990	USA 1989	USA 1990
Australia		2		4	1	6		
Canada	1	1	3	2	3	2	18	12
European Community	34	39			3	3	5	2
Norway	33	22	46	39	8	8		
S Africa	19	20	13	13	10	8	2	
Sweden	3	3	5	2	2	2		
United States					2	1		
Yugoslavia			5	5				2
Argentina							16	3
Brazil			15	23	22	21	36	44
China	6	5	12	9	48	49	17	33
Hong Kong	2	4		1	1		5	3
Others	2	4	1	2			1	1
Ferrosilicon								
Australia						1		1
Canada			1		2	1	15	16
European Community		9			3	2	4	2
Iceland			2	3	5	4	7	
Norway			61	42	19	20	7	11
S Africa	n.a		1	1	4	1		
Sweden			2	1				
Yugoslavia			7	5	4	5	2	
USSR			13	5	3	3	19	18
Bulgaria			2	1				
Polamd				4				
USA					2			

SILICON

	UK 1989	UK 1990	EC(12) 1989	EC(12) 1990	Japan 1989	Japan 1990	USA 1989	USA 1990
China			2	1	19	32	1	2
Argentina					1		5	4
Brazil			5	6	34	25	25	29
Egypt			2	2	1	1		2
Venezuela					2	3	15	13
Others and unidentified		91	2	29	1	2		2

Net Exports
('000 tonnes)

	UK 1989	UK 1990	EC(12) 1989	EC(12) 1990	Japan 1989	Japan 1990	USA 1989	USA 1990
Silicon metal	0.8	1.0	5.8(a)	4.4(a)	0.6	0.7	5.0	9.0
Ferrosilicon	2.1	1.7	26.1	15.9	2.8	3.8	49.1	50.1

(a) Excluding Germany, Denmark and France

Consumption
('000 tonnes)

	UK 1989	UK 1990	EC(12) 1989	EC(12) 1990	Japan 1989	Japan 1990	USA 1989	USA 1990
Silicon metal	c.42	c.44	c.228	c.229	135	134	181	198 (reported)
Ferrosilicon	41	36 (iron & steel industry)	c.662	c.756 (apparent)	c.479	c.533 (apparent)	327	447 (reported)
Total (Si content)							548	588 (apparent)

Import Dependence
Imports as % of consumption:

	UK 1989	UK 1990	EC(12) 1989	EC(12) 1990	Japan 1989	Japan 1990	USA 1989	USA 1990
Silicon metal	100	100	58	59	100	100	24	34
Ferrosilicon	100	100	40	50	85	89	54	54

Imports as % of consumption
and net exports:

	UK 1989	UK 1990	EC(12) 1989	EC(12) 1990	Japan 1989	Japan 1990	USA 1989	USA 1990
Silicon metal	100	100	57	58	99	99	23	32
Ferrosilicon	100	100	39	49	84	88	47	48

Share of World Consumption (%)
Total world (approx. based on production figures)

	UK 1989	UK 1990	EC(12) 1989	EC(12) 1990	Japan 1989	Japan 1990	USA 1989	USA 1990
Silicon metal	5	5	27	26	16	15	22	23
Ferrosilicon	1	1	13	15	9	11	6	9

Consumption growth (% p.a.)

	UK	EC(12)	Japan	USA
1970s total	-2	n.a	5.8	2.4
of which:Silicon metal	2.9	n.a	10.3	6.9
Ferrosilicon	-5.2	2 (approx.)	5	0.8
1980s total	-1.4	0.4	2.8	1.3
of which:Silicon metal	6.1	2.6	5.9	5.4
Ferrosilicon	-8.6	-0.1	2.2	-0.1

SILVER

WORLD RESERVES
('000 tonnes and % of total)

Developed			Developing			Centrally Planned		
Australia	24	(8.6)	Mexico	37	(13.2)	USSR	44	(15.7)
Canada	37	(13.2)	Peru	25	(8.9)	Others	12	(4.3)
USA	31	(11.1)	Others	40	(14.3)			
Others	30	(10.7)						
Totals	**122**	**43.6**		**102**	**(36.4)**		**56**	**(20.0)**
Grand Total		**280**						

The reserve base is approximately 420,000 tonnes.

Identified world resources are estimated at 770,000 tonnes. The greater part of reserves and resources is associated with base metals such as copper, lead and zinc.

WORLD MINE PRODUCTION, 1989-90
(tonnes of metal and % of total 1990)

Developed	1989	1990	% 1990	Developing	1989	1990	% 1990	Centrally Planned	1989	1990	% 1990
Australia	1079	1273	(8.4)	Bolivia	267	280	(1.9)	Poland	1003	1000	(6.6)
Canada	1262	1380	(9.1))	Chile	545	633	(4.2)	USSR	1520	1400	9.3)
Japan	156	150	(0.9)	Mexico	2306	2346	(15.5)	Others	267	263	(1.7)
S Africa	182	161	(1.1)	Morocco	237	236	(1.6)				
Spain	450	500	(3.3)	Peru	1840	1725	(11.4)				
Sweden	228	220	(1.5)	Others	1035	1017	(6.7)				
USA	2007	2170	(14.4)								
Yugoslavia	133	105	(0.7)(a)								
Others	243	249	(1.7))								
Totals	**5740**	**6208**	**(41.1)**		**6230**	**6237**	**(41.3)**		**2790**	**2663**	**(17.6)**
Grand Totals	**1989-**	**14760**									
	1990-	**15108**									

(a) Smelter production.

SILVER

PRODUCTIVE CAPACITY, 1990 (major producers)
(tonnes of metal)

Developed		Developing		Centrally Planned	
Australia	1120	Chile	520	Poland	1060
Canada	1380	Mexico	2420	USSR	1500
Japan	360	Peru	2050		
Spain	370				
USA	2130				
Totals	**5360**		**4990**		**2560**

The above countries account for 81% of world capacity (c. 15880 tonnes).

WESTERN WORLD SILVER SUPPLIES
(tonnes of metal)

	1988	1989	1990
Mine production	11004	11309	11415
Other sources of supply			
US Treasury & Strategic Stockpile Sales	243	411	342
Other governments sales/(purchases)	(168)	(628)	56
Secondary sources incl. demonetised coins	2958	3029	2902
Liquidation of (additions to)			
private bullion stocks(a)	(78)	(1138)	(566)
Net trade with Centrally			
Planned economies(b)	510	778	834
Available for Western world consumption(c)	14469	13761	14983

Source: The Silver Market 1990, Handy & Harman

(a) Including changes in exchange stocks.
(b) Total USSR & E. Europe production less consumption.
(c) Before allowing for Indian stock changes.

RESERVE/PRODUCTION RATIOS

Static Reserve Life (years):	18.5
Ratio of identified reserve base to	
cumulative primary demand 1991-2010:	2.2 : 1

This ignores substantial secondary recovery and above ground stocks.

CONSUMPTION

	tonnes		% p.a. growth rates	
	1989	1990	1970s	1980s
Industrial uses				
European Community (a)	3496	3708	-1.9	-0.3
Japan	3176	3207	3.9	4.5
United States	3695	3583	0.8	-1.7
Other Countries	3570	3579	5.2	7.5
Total Western World	**13937**	**14077**	**1.1**	**1.7**
Coinage (b)	1082	793	-2.6	1.5
Total Consumption				
(W World)	**15019**	**14870**	**0.8**	**1.7**
USSR & Eastern Europe	2952	2865	n.a	n.a.

Source: Handy & Harman reports on silver

(a) France, W Germany, Italy and United Kingdom only.
(b) Demand for coinage is highly volatile as it depends on government programmes in only a few major using countries.

END USE PATTERNS ,1990 (%)

This covers only industrial uses and excludes 'investment' demand.

USA		Japan	
Photography	50	Photography	52
Electrical and electronic components	25	Nitrates	9
Sterlingware and electroplated ware*	10	Contacts	10
Brazing alloys and solders	5	Solder & brazing	4
Others	10	Plating	4
		Silverware	4
		Others	17

* A higher percentage in less industrialised countries.

VALUE OF CONTAINED METAL IN ANNUAL PRODUCTION

$2.0 billion (mine production at average 1991 prices).

SUBSTITUTES

Stainless steel is an economic alternative in table flatware. Aluminium and rhodium are used for reflecting surfaces. Tantalum is a substitute for surgical plates, pins and sutures.

Silver has been replaced in coinage in many countries by cupro-nickel, cupro-zinc, nickel and aluminium.

SILVER

Gold or platinum group metals can be substituted for silver in electrical and electronic components, increasing resistance to oxidation. Silverless black and white film and xerography have replaced silver-containing films in some applications. Video and ultrasonic scanning threaten silver based photographic film.

TECHNICAL POSSIBILITIES

Improvements in solid-state switching and in electroplating and cladding technology will extend life of electronic equipment, decreasing demand. Replacement of silver batteries by lithium batteries. Development of non-silver brazing alloys. Fears over possible environmental/health hazards could restrict some uses.

PRICES

	1986	1987	1988	1989	1990	1991
LME Cash/London Bullion market (a) $/troy oz	5.44	7.02	6.50	5.49	4.83	4.05
LME Cash Real Dec 1991 prices	6.32	7.93	7.09	5.71	4.83	4.03
LME Cash Monthly range $/troy oz	5.02-6.03	5.48-8.46	6.10-7.07	5.13-5.96	3.95-5.36	3.55-4.57

(a) London Bullion market from mid 1989.

Prices result from the interaction of supply and demand with variable, and sometimes considerable, speculative activity. The London Metal Exchange withdrew its silver contract in mid 1989. The London Bullion market quotes silver, as do exchanges in New York and Chicago.

MARKETING ARRANGEMENTS

A fairly small share of newly mined silver is from predominantly silver ores, most being derived as by-product of copper, lead and zinc. Demand exceeds primary supply and the deficit is supplied by secondary sources of various types, and from abundant above-ground stocks. At the end of 1990 total government stocks were estimated at 8,400 tonnes and private stocks at nearly 40,000 tonnes. Of this latter total 8,650 tonnes was held on Commodity Exchanges (predominantly the New York Commodity Exchange).

REAL PRICES 1979 to 1991
Silver, LME/London Bullion Market

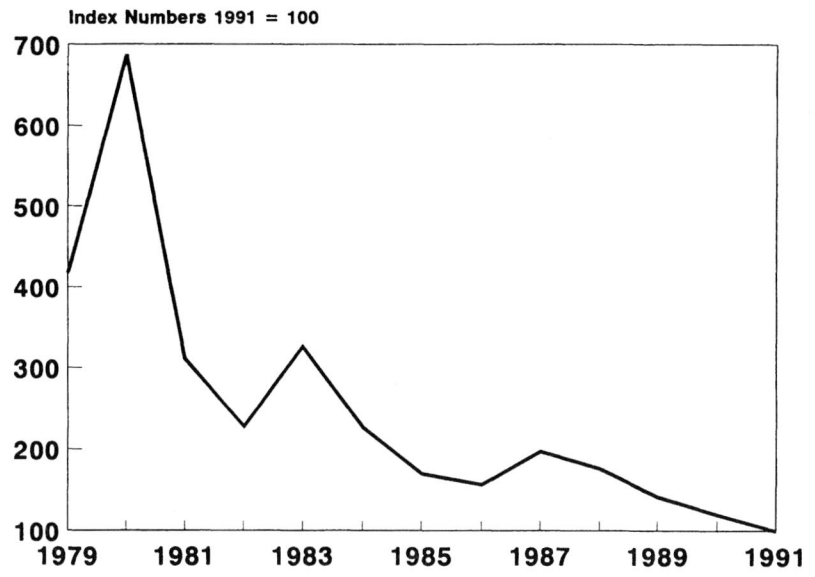

Index Numbers 1991 = 100

WORLD PRODUCTION 1979 to 1991
Silver

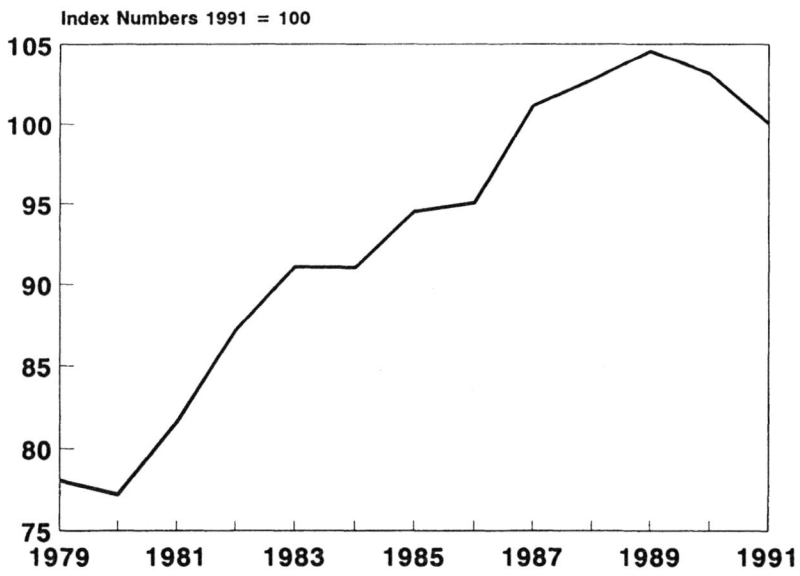

Index Numbers 1991 = 100

SILVER

SUPPLY AND DEMAND BY MAIN MARKET AREA

	UK		EC(12)		Japan		USA	
	1989	1990	1989	1990	1989	1990	1989	1990
Production (tonnes)								
Mine production	1.7	2.7	459	485.5	155.8	149.9	2007	2068
Refinery production:								
New	n.a	n.a	n.a	n.a	-	-	1718	1611
Secondary (old scrap)	330	364	1269(a)	1275(a)	218	227	735	433
Net Imports (tonnes)								
Ores and concentrates	6	26	282	1085	9142	6390	125.2(a)	86.6(b)
Unwrought (inc. powders)	1003	2561	3839	4736	853	932	3219.9	2762.0
Wrought (inc. partly worked)	207	153	406	366	377	557	3.1	2.8
Waste and scrap	n.a	n.a	n.a	n.a	38	112	n.a	n.a

(a) Includes scrap of platinum group metals.
(b) Silver content.

Source of Net Imports (%)

Metal	UK 1989	UK 1990	EC(12) 1989	EC(12) 1990	Japan 1989	Japan 1990	USA 1989	USA 1990
Australia	5	2	2	1	6	5		
Austria								
Canada	2		1	1		1	36	39
European Community	14	9			3	3		1
Finland	2							
S Africa	11		3					
Sweden			5	3				
Switzerland			12	23				
United States	13	3	21	12	10	25		
Yugoslavia			1					
E Germany	24	6	6	3				
N Korea								
Poland			4	3				
Chile			3	2	3	1	1	2
Hong Kong								
Mexico	18	64	7	36	49	44	55	45
Morocco			30	3				
Peru					23	18	3	9
Philippines					1			
S Korea					3	1		
Zambia	2							
Namibia		10		5				
Uruguay							2	3
Singapore			1		2			
Others and unspecified	9	6	4	8		2	2	1

	UK		EC(12)		Japan		USA	
	1989	1990	1989	1990	1989	1990	1989	1990
Net Exports								
(tonnes)								
Ores and concentrates	1	-	54	27	-	-	0.6	21.9
Unwrought (inc. powders)	1078	2955	1768	3077	6.0	5.4	602.5	789.4
Wrought (inc. partly worked)	178	155	1431	681	576	14.8	77.4	72.1
Waste/scrap	n.a	n.a	n.a	n.a	59.0	12.0	n.a	n.a

(a) includes scrap of platinum group metals.

	UK		EC(12)		Japan		USA	
Consumption								
(tonnes)								
Industrial	700	722	c.4500	c.4750	3176	3207	3695	3583
Coinage			361(a)	118(a)			292	236

(a) France and Germany only.

Import Dependence
Imports as % of consumption

(exc. coinage)	100	100	64	63	88	88	26	30

(based on mine and scrap production)

Changes in unreported stocks preclude effective analysis of overall market shares of imports.

Share of World Industrial Consumption (%)

	UK		EC(12)		Japan		USA	
Western World	5	5	c.32	c.32	23	22	26	24
Total World	4	4	c.27	c.28	19	19	22	21

Consumption Growth (% p.a.)

	UK	EC(12)	Japan	USA
1970s	-0.7	-1.9	3.9	0.8
1980s	-0.1	-0.3	4.5	-1.7

Note: Some of the figures in this table (e.g. on consumption) differ from those of earlier tables, derived from different sources.

SULPHUR

WORLD RESERVES
(million tonnes and % of total)

Developed			Developing			Centrally Planned		
Canada	158	(11.3)	Iraq	130	(9.3)	China	100	(7.1'
France	10	(0.7)	Mexico	75	(5.4)	Poland	130	(9.3'
W Germany	17	(1.2)	Saudi Arabia	100	(7.1)	USSR	250	(17.9'
Italy	10	(0.7)	Others	125	(8.9)	Others	60	(4.3'
Japan	5	(0.3)						
Spain	50	(3.6)						
USA	140	(10.0)						
Others	40	(2.9)						
Totals	430	(30.7)		430	(30.7)		540	(38.6'
Grand Total		1400						

The reserve base is 3,500 million tonnes with major deposits located in Canada, Iraq, Spain, Poland and USSR. Identified world resources total 5,000 million tonnes.

WORLD PRODUCTION IN ALL FORMS, 1989-90
('000 tonnes and % of total 1990)

A significant percentage of output is a by-product of metallurgical operations or petroleum refining.

Developed	1989	1990	% 1990	Developing	1989	1990	% 1990	Centrally Planned	1989	1990	% 1990
Canada	6697	6947	(11.9)	Iraq	1270	1050	(1.8)	China	4900	4900	(8.4)
Finland	636	632	(1.1)	Mexico	2369	2340	(4.0)	E Germany	290	265	(0.5)
France	1036	1045	(1.2)	Saudi Arabia	1400	1500	(2.6)	N Korea	230	230	(0.4)
W Germany	1885	1835	(3.2)	Others	3446	3318	(5.7)	Poland	5150	5030	(8.7)
Italy	510	490	(0.8)					Romania	270	230	(0.4)
Japan	2623	2703	(4.7)					USSR	9900	9025	(15.5)
S Africa	685	688	(1.2)					Others	112	112	(0.2)
Spain	1058	1085	(1.9)								
Sweden	398	400	(0.7)								
USA	11592	11560	(19.9)								
Yugoslavia	428	423	(0.7)								
Others	1833	2313	(3.9)								
Totals	29381	30121	(51.8)		8485	8208	(14.1)		20852	19792	(34.1)
Grand Totals	1989-	58718									
	1990-	58121									

Of the total output 19% was Frasch, 5% was native sulphur and 17% was from pyrites. The balance came from by-product sources.

PRODUCTIVE CAPACITY, 1990
('000 tonnes)

Developed		Developing		Centrally Planned	
Canada	8400	Iraq	1600	China	5450
France	1850	Mexico	2775	Poland	4900
W Germany	2520	Saudi Arabia	1780	USSR	11000
Italy	830	Others	4125	Others	1000
Japan	4100				
S Africa	1000				
Spain	1550				
USA	12300				
Yugoslavia	750				
Others	4090				
Totals	37400		10280		22350
Grand Total	70030				

RESERVE/PRODUCTION RATIOS

Static ReserveLlife (years):	24
Ratio of identified reserve base to cumulative demand 1991-2010:	2.9 : 1

CONSUMPTION

	Sulphur in all forms '000 tonnes		% p.a. growth rates	
	1989	1990	1970s	1980s
European Community(a)	c.7150	c.6815	0.7	-1.7
Japan	2715	2775	-1.4	1.2
United States	12685	13056	3.4	-0.6
Other Countries(a)	17571	18870	5.5	3.0
Total Western World	**40121**	**41516**	**3.0**	**0.8**
Centrally Planned Economies	20985	19668	n.a	1.8
Total World	**61106**	**61184**	**n.a**	**1.1**

Source: British Sulphur Corporation statistics and trade accounts.
(a) Definitions may not be consistent during 1980s.

END USE PATTERNS 1990 (USA) (%)

Fertilisers	66
Other chemical products	14
Metal mining	5
Petroleum refining	5
Other uses	10

SULPHUR

VALUE OF ANNUAL PRODUCTION

$5.1 billion (at average 1991 fob prices).

SUBSTITUTES

Most sulphur is used in the form of sulphuric acid. Depending on relative prices, this can sometimes be replaced by hydrochloric acid, nitric acid or hydrofluoric acid.

TECHNICAL POSSIBILITIES

Sulphur-asphalt paving, a non-asphalt paving with a sulphur binding, and specialised sulphur concrete materials are near commercial use.

PRICES

	1986	1987	1988	1989	1990	1991
US Frasch, liquid bright, fob Holland $/lt	152.5	137.4	120.0	120.8	121.7	126.8
Real Dec 1991 prices	177.1	155.3	130.9	125.8	122.1	126.4
Liquid sulphur contracts North West Europe delivered, ex-terminal range $/tonne	149.5-164.5	120-140	120-124	122-133	116-121	105.8-133.5

Producer pricing for long term contracts, although spot market is important. Transport costs are very important. Thus in 1991 posted prices changed every six months. The fob prices in the USA were between $70 and $105/tonne versus the $106-133/tonne ex-terminal in North West Europe.

MARKETING ARRANGEMENTS

Approximately 50% of world production is from countries in which the industry is nationalised (eg: USSR, Mexico) or in which the governments have partial ownership (eg: France, Spain) or exercise some measure of control (eg: Japan). Production is worldwide with elemental sulphur (frasch and native) accounting for about 25% of Western world primary production, pyrites 17%, smelter gases 14% and natural gas or oil most of the balancing 44%. The supply pattern is likely to be restructured by 2000, as production from co-product sources expands(coal, petroleum, natural gas, metal smelting) in response to environmental pressures.

REAL PRICES 1979 to 1991
Sulphur, US Frasch fob Rotterdam

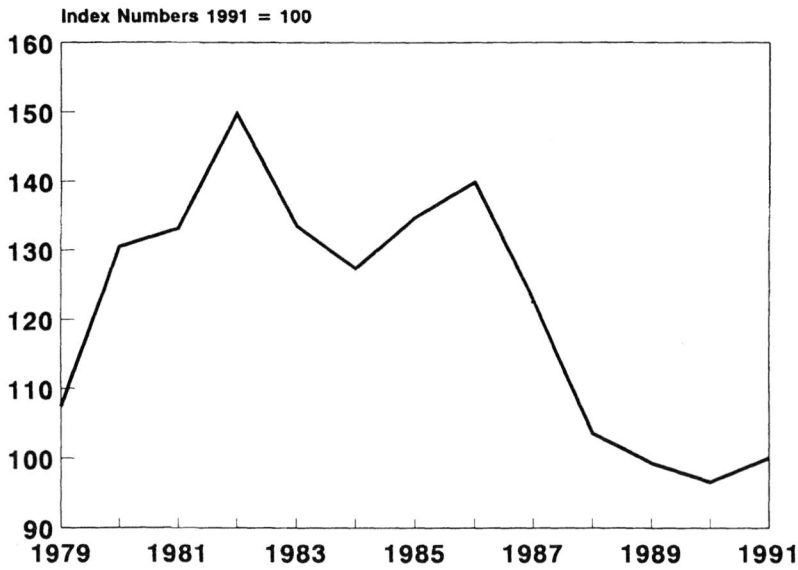

WORLD PRODUCTION 1979 to 1991
Sulphur

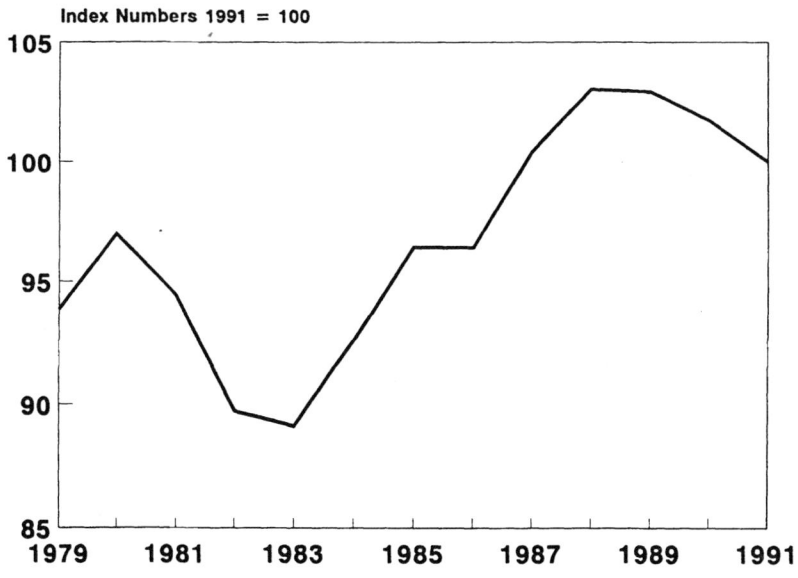

SULPHUR

SUPPLY AND DEMAND BY MAIN MARKET AREA

	UK		EC(12)		Japan		USA	
	1989	1990	1989	1990	1989	1990	1989	1990
Production ('000 tonnes)								
Sulphur in all forms	185	185	6155	5915	2656	2842	11592	11560
Net Imports ('000 tonnes)	650	485	1298	1371	328	318	2260	2571
Source of Net Imports (%)								
Australia					22	23		
Canada			14	20			52	55
European Community							1	1
Finland			8	8	6	5		
Sweden			3	3				
Norway			2	2				
Yugoslavia			2	3				
USSR	n.a	n.a	8	7		6		
United States				17	14			
China					71	64		
Poland			33	33				
Iraq			3	4				
Mexico							47	43
Saudi Arabia	..	..	9	4				
Others	..	..	4	2	1	2		1
Net Exports ('000 tonnes)	4.0	4.1	321	475	271	386	1024	972
Consumption ('000 tonnes)								
Sulphur in all forms	830	666	c.7150	c.6815	2715	2775	12685	13056
Import Dependence								
Imports as % of consumption	78	73	18	20	12	12	18	20
Imports as % of consumption and net exports	78	72	17	19	11	10	16	18
Share of World Consumption (%)								
Western world	2	2	18	16	7	7	32	31
Total world	1	1	12	11	4	5	21	21
Consumption Growth (% p.a.)								
All forms 1970s	-0.7		0.7		-1.4		3.4	
1980s	-6.0		-1.7		1.2		-0.6	

TALC

WORLD RESERVES
(million tonnes and % of total)

Developed			Developing			Centrally Planned	
Japan	4.0	(1.1)	Brazil	14.0	(3.7)	China	large
USA	132.0	(35.0)	India	4.0	(1.1)	USSR	large
Others	135.0	(35.8)	S.Korea	14.0	(3.7)		
			Others	74.0	(19.2)		
Totals	271.0	(71.9)		106.0	(28.1)		large
Grand Total(Western World)		377.0					

The reserve base is estimated at 1.12 billion tonnes. These figures cover reserves of talc and related minerals, including pyrophyllite.

WORLD MINE PRODUCTION, 1989-90
(millions of tonnes and % of total 1990)
Talc and Related Minerals

Developed	1989	1990	% 1990	Developing	1989	1990	% 1990	Centrally Planned	1989	1990	% 1990
Australia	200.0	205.0	(2.5)	Argentina	28.2	27.6	(0.3)	China	1000.0	1000.0	(12.1)
Austria	133.0	130.0	(1.5)	Brazil	715.0	720.0	(8.7)	N.Korea	170.0	170.0	(2.0)
Canada	146.0	146.0	(1.8)	India	511.5	511.0	(6.2)	Romania	60.0	55.0	(0.7)
Finland	380.0	380.0	(4.6)	Iran	30.0	30.0	(0.4)	USSR	530.0	500.0	(6.1)
France	329.9	330.0	(4.0)	S. Korea	932.4	910.0	(11.0)	Others	12.0	12.0	(0.1)
W.Germany	13.0	15.0	(0.2)	Mexico	13.9	13.8	(0.2)				
Greece	10.5	10.0	(0.1)	Pakistan	38.3	40.0	(0.5)				
Italy	146.0	145.0	(1.8)	Taiwan	22.5	22.0	(0.3				
Japan	1289.3	1292.7	(15.7)	Thailand	47.0	46.2	(0.6)				
Norway	100.0	100.0	(1.2)	Uruguay	1.6	16.0	(0.2)				
Portugal	8.1	7.9	(0.1)	Others	27.4	28.2	(0.3)				
S.Africa	15.5	13.9	(0.2)								
Spain	75.0	75.0	(0.9)								
Sweden	16.0	16.0	(0.2)								
UK	15.4	15.0	(0.2)								
USA	1253.2	1267.2	(15.3)								
Totals	4130.9	4148.7	(50.3)		2367.8	2364.8	(28.7)		1772.0	1737.0	(21.0)
Grand Totals	1989- 1990-	8270.7 8250.5									

of which:	1989	1990
Pyrophyllite	2433.5	2417.0
Steatite	633.1	631.3
Talc	3590.0	3588.3
Unspecified	1614.5	1614.2

TALC

PRODUCTIVE CAPACITY, 1990
Talc and Pyrophyllite ('000 tonnes)

Developed		Developing		Centrally Planned	
Australia	227	Argentina	45	China	1179
Austria	136	Brazil	454	USSR	544
Canada	299	India	454	Others	1
Finland	363	Iran	32		
France	363	Mexico	36		
Italy	163	S. Korea	998		
Japan	1633	Others	132		
Norway	136				
S.Africa	18				
USA	1361				
Others	236				
Totals	**4935**		**2151**		**1724**
Grand Total	**8810**				

RESERVE/PRODUCTION RATIOS

Static Reserve Life (years):	Over 46
Ratio of identified reserve base to cumulative demand 1991-2010:	6:1

CONSUMPTION
(Apparent; talc only)

	'000 tonnes		% p.a. growth rates	
	1989	1990	1970s	1980s
European Community	c.750	c.765	n.a	n.a
Japan	c.810	c.725	n.a.	n.a
United States	933	974	n.a	1.0

END USE PATTERNS, 1990(%)

USA		Japan	
Ceramics	28	Paper	86
Cosmetics	4	Plastics	9
Paint	18	Paint	5
Paper	17	Others	..
Plastics	6		
Roofing	11		
Others	16		

VALUE OF ANNUAL PRODUCTION

$1 billion approximately, at average 1991 prices (talc and pyrophyllite).

SUBSTITUTES

Clay and pyrophyllite can be used as substitutes for talc in ceramics; calcium carbonate and kaolin in paint, paper and rubber; clays, feldspar, and mica in plastics.

TECHNICAL POSSIBILITIES

Ceramics, paint and paper continue to provide the major uses for talc.

PRICES

	1986	1987	1988	1989	1990	1991
Norwegian Ground (ex store) UK £/tonne	95	95	95	95.84	100	115
Chinese, normal (ex store) UK 200 mesh £/tonne	140.3	141	141	141.5	144	144
Norwegian Ground UK $/tonne						
Real Dec 1991 prices	161.8	175.9	184.8	163.3	178.9	202.6

MARKETING ARRANGEMENTS

A fragmented industry based on small local deposits has been heavily rationalised in recent years in the hands of a few dominant producers. RTZ (via Talc de Luzenac) is the largest. Prices vary widely according to technical specifications and end uses.

REAL PRICES 1979 to 1991
Talc, Norwegian Ground

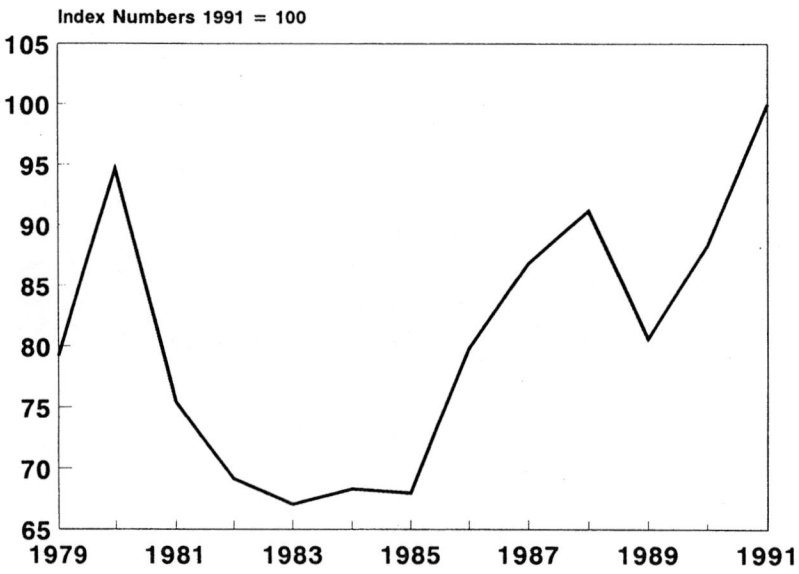

Index Numbers 1991 = 100

WORLD PRODUCTION 1979 to 1991
Talc, including Pyrophyllite

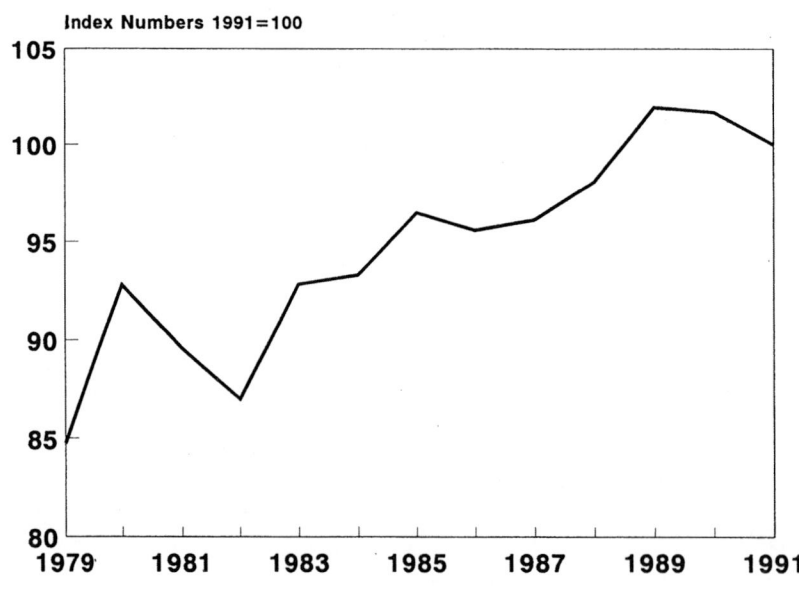

Index Numbers 1991=100

SUPPLY AND DEMAND BY MAIN MARKET AREA

	UK		EC(12)		Japan		USA	
	1989	**1990**	**1989**	**1990**	**1989**	**1990**	**1989**	**1990**
Production								
('000 tonnes)								
Talc	15.4	14.8	512.5	513	55.7	61.6	1171.9	1185.1
Pyrophyllite and other minerals	-	-	85.5	85	1233.6	1231.2	81.3	82.0
Net Imports								
('000 tonnes)	73.6	83.4	291	301	752	664	77	63
Source of Net Imports (%)								
Net Imports (%)								
European Community		5					9	5
China			10	9	79	79	7	10
Australia		1	9	4	17	16	17	20
USA			15	19	3	4		
S.Korea					1		2	
Norway			8	7				
Sweden			5	3				
Finland			20	24				
Austria			31	32				
India			1	1				
Canada							61	63
Japan							3	1
Others and unspecified		94	1	1			1	
Net Exports								
('000 tonnes)	1.8	1.8	5.3	48	-	-	319	200
Consumption								
('000 tonnes)								
Talc	87	96	c.750	c.765	c.810	c.725	933 (apparent)	974
Pyrophyllite, etc.	n.a	n.a	n.a	n.a	n.a	n.a	79 (apparent)	80
Import Dependence								
Imports as % of consumption	85	87	39	39	93	92	8	7
Imports as % of consumption and net exports	83	85	36	37	93	92	6	5
Share of World Consumption (%)								
(Talc only)								
Total World	2	2	14	15	16	14	18	19
Consumption Growth (% p.a.)								
1970s	n.a		n.a		n.a		2.1	
1980s	n.a		n.a		n.a		1.0	

TANTALUM

WORLD RESERVES
(tonnes of metal and % of total)

Developed			Developing			Centrally Planned		
Australia	4535	(17.6)	Brazil	900	(3.5)	USSR	c.4500	(17.4
Canada	1815	(7.0)	Malaysia	450	(1.7)			
Others	1360	(5.3)	Nigeria	3175	(12.3)			
			Thailand	7250	(28.1)			
			Zaire	1815	(7.0)			
Totals	7710	(29.9)		13590	(52.7)		c.4500	(17.4
Grand Total		25800						

The Western World's reserve base is estimated at 35,000 tonnes, so that the total is some 40,000 tonnes.

WORLD MINE PRODUCTION, 1989-90
(tonnes of contained metal)

Developed	1989	1990	Developing	1989	1990	Centrally Planned	
Australia	119	119	Brazil	127	108	Total	n.a
Spain	3	3	Namibia	..	1		
Canada	73	82	Nigeria	3	2		
			Thailand	29	2		
			Zaire	14	14		
			Zimbabwe	11	12		
Totals	195	204		181	139		n.a

TOTALS
(Western World only)   1989-   376
                       1990-   343

Tantalum Mining Corporation's Canadian mine, the Australian Greenbushes mine of Gwalia Consolidated, and the Wodgina mine of Pan West Tantalum, which opened in 1990, are the three main primary sources.

Malaysia, Zambia and the United States produced minor tonnages.

This table excludes production of tantalum from tin slags which is concentrated in SE Asia, Africa and the USSR (greater than 100), with other countries producing small quantities. Excluding the USSR, world tantalum production from tin slags was 362 tonnes in 1989 and 343 tonnes in 1990. This means that total Western World production of tantalum from all sources was roughly 738 tonnes in 1989 and 686 tonnes in 1990. Total world production, including the USSR and China, is close to 900 tonnes/year.

Because tin slags are excluded, no percentage shares of world output are given in the table.

The production of the members of the Tantalum Producers' International Study Centre (TIC) was as follows:

(tonnes contained Ta$_2$O$_5$)	1988	1989	1990
Production			
Slags	634	442	419
Concentrates	134	300	320
Total	**768**	**742**	**739**

Source: TIC Quarterly Bulletins.

Processors' shipments amounted to 932 tonnes of contained tantalum in 1989 and 948 tonnes in 1990. These include materials from secondary as well as primary sources.

PRODUCTIVE CAPACITY, 1990
(tonnes of contained metal)

Developed		Developing		Centrally Planned	
Australia	180	Brazil	180	USSR	n.a
Canada	90	Malaysia	90		
Spain	c.5	Nigeria	45		
		Thailand	365		
		Zaire	45		
		Other Africa	90		
Totals	**275**		**815**		**n.a.**
Total (W.World only)	**1000**				

This table includes estimates of capacities of by-product recovery from tin slags.

RESERVE/PRODUCTION RATIOS

Static Reserve Life (years):	75
Ratio of identified reserve base to cumulative demand 1991-2010:	1.4 : 1

CONSUMPTION

	Tonnes		% p.a. growth rates	
	1989	1990	1970s	1980s
European Community	c.225	c.225	n.a	n.a
Japan	c.300	c.300	13.6	n.a
			(powder only)	
United States	376	390	1.8	-3.2

TANTALUM

END USE PATTERNS, 1990(%)

USA		Japan		Western World	
Electronic components	60	Electronics	60	Capacitators	46
Machinery	11	Industrial	10	Carbide	25
Transport	15	Cutting tools	20	Mill Products	15
Others	14	Others	10	Others	14

VALUE OF CONTAINED METAL IN ANNUAL PRODUCTION

$45 million (at average 1991 prices) (including tantalum content of tin slags).

SUBSTITUTES

Substitution for tantalum is normally at the expense of performance or cost.

Aluminium and ceramics compete in capacitors; silicon, germanium and selenium are alternatives in rectifiers; zirconium, titanium can substitute as getters in electronic tubes and in corrosion-resistant equipment.

Niobium can replace tantalum in some carbides, and, along with platinum, in corrosion-resistant equipment and high temperature uses.

Hafnium, molybdenum and tungsten compete in high temperature applications.

TECHNICAL POSSIBILITIES

Reduction in quantity of tantalum required per capacitor through higher capacitance ratings, and miniaturisation of components.

Increased use in superalloys and development of new alloys.

PRICES

	1986	1987	1988	1989	1990	1991
Tantalite Ore 60% Ta_2O_5 spot, cif US ports						
$/lb Ta_2O_5	19.5	22.5	38.8	36.8	31.2	30.0
Real Dec 1991 prices	22.6	25.4	42.3	38.3	31.2	29.9

Tanco's list price was suspended in early 1985. The dealer market is important, although most producers sell direct to customers.

MARKETING ARRANGEMENTS

Most tantalum is produced in association with tin or niobium. Developments in these products greatly influence production. Production divided between mining of ores, eg Australia and Canada, and processing of tin slags, eg Thailand. Some countries combine both methods, eg Brazil. The Tantalum Producers' International Study Centre (TIC) based in Brussels, carries out cooperative statistical, promotional and research activities. Most primary producers and processors are members.

REAL PRICES 1979 to 1991
Tantalum, Tantalite ore cif US ports

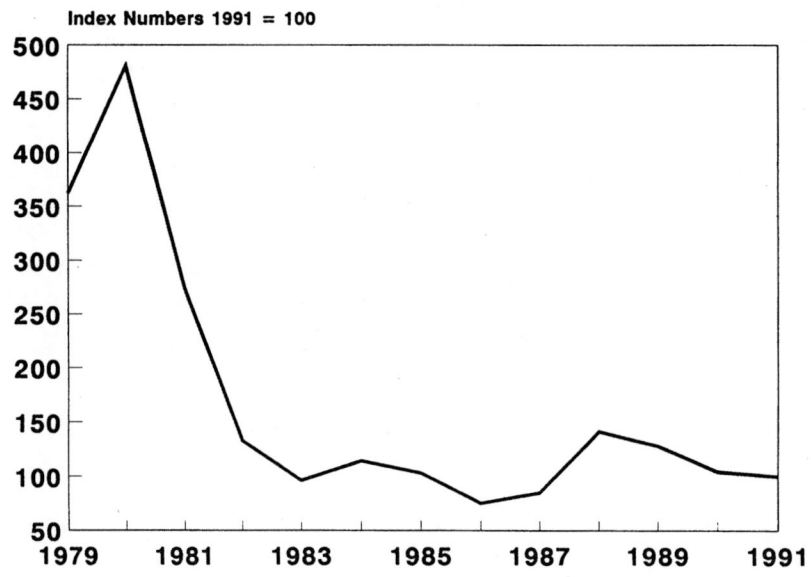

Index Numbers 1991 = 100

WORLD PRODUCTION 1979 to 1991
Tantalum

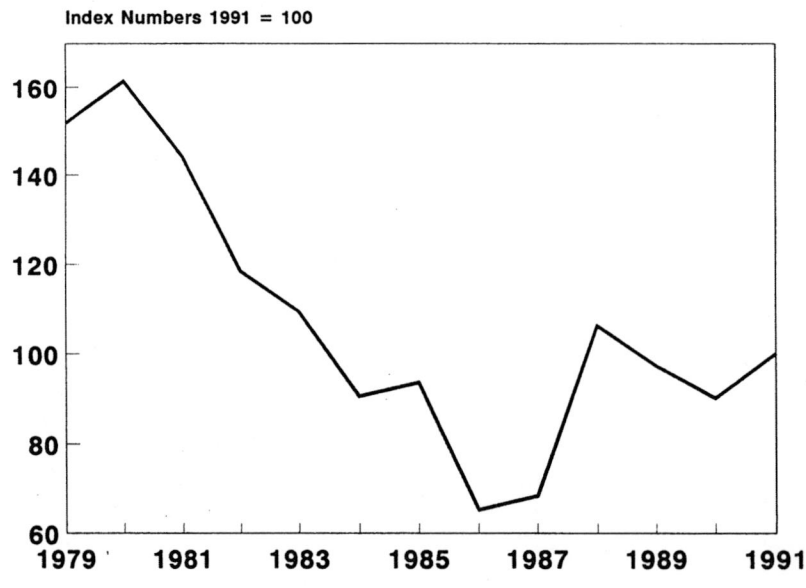

Index Numbers 1991 = 100

SUPPLY AND DEMAND BY MAIN MARKET AREA

	UK 1989	UK 1990	EC(12) 1989	EC(12) 1990	Japan 1989	Japan 1990	USA 1989	USA 1990
Production (tonnes Ta content)								
Ores and concentrates	-	-	3	3	-	-	small	small
Net Imports (tonnes)								
Mineral concentrates	inc. with niobium		inc. with niobium		294(a)	332(a)	1787	1047
Unwrought metal and alloys inc. powder	48	57	182(b)	204(b)	39	28	109	95
Wrought metal	26	24	51	54	25	27	1	3

(a) Includes vanadium ores and concentrates.
(b) Excluding Belgium/Luxembourg.

Source of Net Imports (%)

Ores and Concentrates

	UK 1989	UK 1990	EC(12) 1989	EC(12) 1990	Japan 1989	Japan 1990	USA 1989	USA 1990
Australia						2	36	72
Canada							4	11
European Community							72	52
Nigeria					3	7		
Brazil	n.a		n.a			7	6	15
Malaysia					18	64		
Burundi						4		
Thailand					49		3	20
Zaire					23	6	3	7
Zimbabwe					2	6		2
Japan							6	3
Others					5	4	3	

Metal (unwrought)

	UK 1989	UK 1990	EC(12) 1989	EC(12) 1990	Japan 1989	Japan 1990	USA 1989	USA 1990
Australia			6				13	2
Austria		12		16				2
European Community	63	40			35	20	52	64
Japan		7	5	6			10	1
United States	33	40	78	69	58	66		
Taiwan			3		1	13	6	6
China			7		5	1		3
Canada							1	3
Mexico							8	12
Hong Kong								5
Others	4	1	1	9	1		10	3

TANTALUM

	UK 1989	UK 1990	EC(12) 1989	EC(12) 1990	Japan 1989	Japan 1990	USA 1989	USA 1990
Net Exports (tonnes)								
Concentrates	incl.with niobium		incl. with niobium				7	13
Metal, alloys, waste and scrap	46	38	45 (a)	56 (a)	21.7	27.9	195	172
							(Ta content)	

(a) Excluding Belgium/Luxembourg.

Consumption (tonnes Ta content)	c.36	c.45	c.225	c.325	c.300	c.300	376	390

Import Dependence

	UK 1989	UK 1990	EC(12) 1989	EC(12) 1990	Japan 1989	Japan 1990	USA 1989	USA 1990
Imports as % of consumption	100	100	100	100	100	100	100	100
Imports as % of consumption and net exports	100	100	100	100	100	100	100	100

Share of World Consumption (%)

Western world	c.4	c.5	c.24	c.24	c.32	c.32	c.40	c.41

Consumption Growth (% p.a.)

	UK	EC(12)	Japan	USA
1970s	n.a	n.a	13.6 (powder only)	1.8
1980s	n.a	n.a	n.a	-3.2

TELLURIUM

WORLD RESERVES
(tonnes of contained tellurium and % of total)

Developed			Developing			Centrally Planned		
Australia	500	(2.3)	Chile	5500	(25.4)	Total	2300	(10.6)
Canada	800	(3.7)	Peru	500	(2.3)			
USA	3700	(17.1)	Papua N. Guinea	400	(1.8)			
Others	1000	(4.6)	Philippines	700	(3.2)			
			Zaire	1700	(7.8)			
			Zambia	2000	(9.2)			
			Others	2600	(12.1)			
Totals	6000	(27.6)		13400	(61.8)		2300	(10.6)
Grand Total		21700						

The above data refer to reserves of by-product tellurium contained in copper deposits of economic grade. The reserve base is 37,000 tonnes. Concentrations of tellurium can also be found in lead and gold ores and in coal deposits.

WORLD REFINERY PRODUCTION OF PRIMARY METAL, 1989-90
(tonnes of metal)

Developed	1989	1990	Developing	1989	1990	Centrally Planned	1989	1990
Belgium	60	60	Peru	5	4	USSR	5	5
Canada	9	13	Philippines	3	6			
Japan	51	50						
USA	69	36						
Others(a)	10	10						
Totals	199	169		8	10		5	5
TOTALS (W world only)	1989-	212						
	1990-	184						

(a) Australia and W Germany.

For most countries, production is estimated with a margin of error of up to 20% i.e. world production is in the 200-250 tonne range. The incompleteness of the data explains the lack of percentage shares in the table.

Chile, Zambia, and Zaire may also refine tellurium but details of production are not available. The Philippines started in 1989.

The Japanese copper refineries, Asarco in the United States, and MHO in Belgium are the major producers.

TELLURIUM

WORLD REFINERY CAPACITY, 1990

World refinery capacity is approximately 600 tonnes concentrated in the USA (110 tonnes), Japan (100 tonnes), Canada (60 tonnes), USSR (70 tonnes), Belgium (100 tonnes), Peru (20 tonnes) and Philippines (100 tonnes). Some sources estimate Belgian capacity at 150 tonnes, but this is not borne out by estimated production levels. Much of the nominal capacity is non operational.

RESERVE/PRODUCTION RATIO

Static Reserve Life (years):	102 (Western World only)
Ratio of reserve base to cumulated demand 1991-2010:	7.5 : 1

CONSUMPTION

Apparent consumption based on incomplete data.

	tonnes		% p.a. growth rates	
	1989	1990	1970s	1980s
European Community	c.115	c. 85	n.a	n.a
Japan	c.50	c.50	n.a	n.a
United States	c.110	c.80	1.2	n.a

END USE PATTERNS, 1990 (%)

USA

Iron and steel products	55
Non-ferrous metals	20
Chemicals, including rubber manufactures	17
Xerography and others	8

Western World

Metallurgy	55
Chemicals	25
Electrical	15
Other	5

VALUE OF ANNUAL PRODUCTION

$15 million (identified production at average 1991 prices).

SUBSTITUTES

Bismuth is being increasingly substituted for tellurium in free machining steels, with selenium and lead as other alternatives in metallurgical applications. Selenium and sulphur can be used in rubber compounding applications and selenium and germanium in electronics.

TECHNICAL POSSIBILITIES

Development of photoelectrochemical solar cells and screen-printed thin film cadmium-sulphide/cadmium telluride solar cells. Potential for other photoactive devices.

New tellurium-containing catalysts.

Recovery from coal deposits.

PRICES

	1986	1987	1988	1989	1990	1991
Metal - Major producer USA $/lb 99.7% min (a)	10	14	35	34	31	32
Real Dec 1991 prices	11.6	15.8	38.2	35.4	31	31.9

(a) List prices suspended May 1981. Year end prices quoted by USBM for US producer.

MARKETING ARRANGEMENTS

As tellurium is recovered as a by-product, supply may move independently of demand. There were shortages in 1988-90 of about 10 tonnes/year, kept from stocks. In 1990 the deficit arose from weakening production rather than increased demand.

The Selenium and Tellurium Development Association, supported by primary producers in Canada, Japan, Peru and the USA, underpins research aimed at encouraging new applications.

REAL PRICES 1979 to 1991
Tellurium, Metal US Major Producer

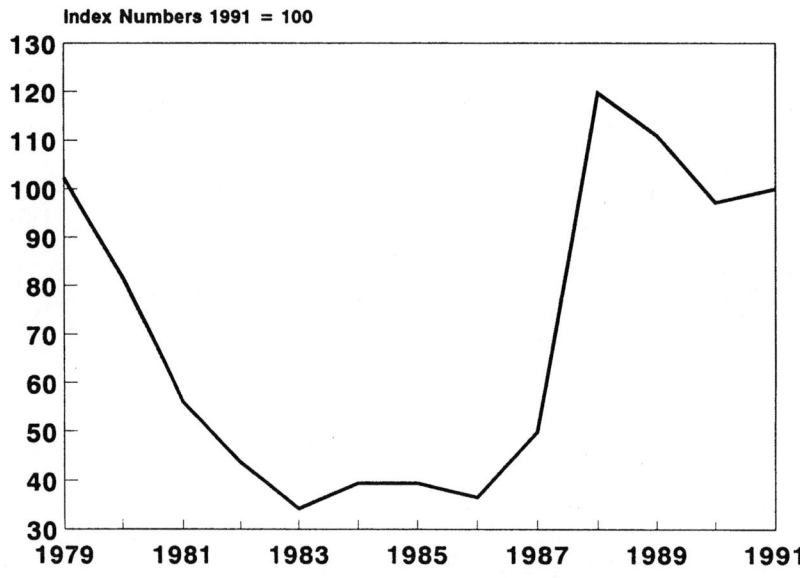

Index Numbers 1991 = 100

WORLD PRODUCTION 1979 to 1991
Tellurium

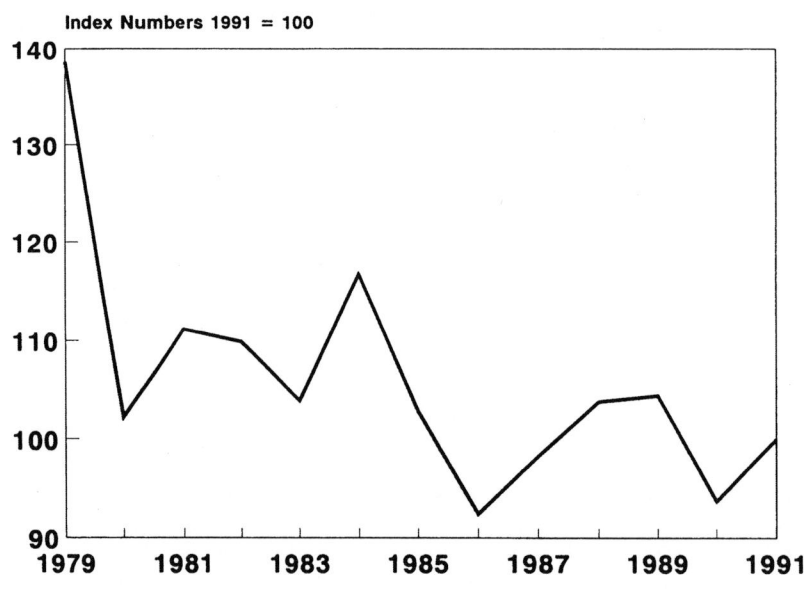

Index Numbers 1991 = 100

SUPPLY AND DEMAND BY MAIN MARKET AREA

	UK		EC(12)		Japan		USA	
	1989	1990	1989	1990	1989	1990	1989	1990
Production (tonnes)								
Refinery	-	-	c.65	c.65	51.0	49.7	69	36
Net Imports (tonnes)	83	40	83	60	4.6(a)	6.3 (a)	42.9	34
								(Te content)

(a) Includes boron.

Source of Net Imports (%)

	UK 1989	UK 1990	EC(12) 1989	EC(12) 1990	Japan 1989	Japan 1990	USA 1989	USA 1990
USSR					44	16		
Canada							25	
European Community	29	53			41	61	50	
Japan			12	n.a.			4	
USA					6	9		
Peru								
Philippines						1	1	
Mexico							2	
China					5			
Others	71	47	88		9	8	18	

	UK 1989	UK 1990	EC(12) 1989	EC(12) 1990	Japan 1989	Japan 1990	USA 1989	USA 1990
Net Exports (tonnes)	27	21	34	42	8(a)	10(a)	n.a	n.a.

(a) Includes boron.

	UK 1989	UK 1990	EC(12) 1989	EC(12) 1990	Japan 1989	Japan 1990	USA 1989	USA 1990
Consumption (tonnes)	n.a	19 (apparent)	c.115	c.85	c.50	c.50	c.110	c.80

Import Dependence (%)

	UK 1989	UK 1990	EC(12) 1989	EC(12) 1990	Japan 1989	Japan 1990	USA 1989	USA 1990
Imports as % of consumption	100	100	c.72	c.71	c.9	c.12	c.39	c.42
Imports as % of consumption and net exports	100	100	c.56	c.47	c.8	c.10	n.a	n.a

Share of World Consumption (%)

	UK 1989	UK 1990	EC(12) 1989	EC(12) 1990	Japan 1989	Japan 1990	USA 1989	USA 1990
Western world	c.18	c.8	c.38	c.34	c.17	c.20	c.37	c.32

Consumption Growth (% p.a.)

	UK		EC(12)		Japan		USA	
1970s	n.a		n.a		n.a		1.2	
1980s	n.a		n.a		n.a		n.a	

TIN

WORLD RESERVES
('000 tonnes of metal and % of total)

Developed			Developing			Centrally Planned		
Australia	200	(3.4)	Bolivia	140	(2.4)	China	1500	(25.4)
Canada	60	(1.0)	Brazil	1200	(20.4)	USSR	-300	(5.1)
Portugal	70	(1.2)	Burma	20	(0.3)	Others	30	(0.5)
S Africa	30	(0.5)	Indonesia	680	(11.5)			
UK	90	(1.5)	Malaysia	1100	(18.6)			
USA	20	(0.3)	Namibia	60	(1.0)			
			Nigeria	20	(0.3)			
			Thailand	270	(4.6)			
			Zaire	20	(0.3)			
			Zimbabwe	20	(0.3)			
			Others	80	(1.4)			
Totals	**470**	**(7.9)**		**3610**	**(61.1)**		**1830**	**(31.0)**
Grand Total		**5910**						

The world's reserve base is 6 million tonnes.

Total identified world resources are estimated at 37 million tonnes.

WORLD MINE AND METAL PRODUCTION, 1989-90, AND PRODUCTIVE CAPACITY, 1990
('000 tonnes of metal and % of total 1990)

	Mine				Metal(a)			
	Production		% of Production	Capacity	Production		% of Production	Capacity
	1989	1990	1990		1989	1990	1990	
Developed								
Australia	7.8	7.4	(3.5)	10	0.4	0.3	(0.1)	2
Canada	2.8	2.8	(1.3)	5	-	-	(-)	-
Japan	-	-	(-)	-	0.8	0.8	(0.4)	2
Netherlands	-	-	(-)	-	4.5	6.1	(2.8)	7
Portugal	0.1	1.4	(0.7)	6	0.1	0.1	(...)	2
S Africa	1.3	1.1	(0.6)	3	2.6	(1.5)	(0.7)	3
Spain	0.1	0.1	(-)	-	1.8	-	-	14
UK	3.8	3.4	(1.6)	5	3.6	6.1	2.8	12
USA	0.1	0.1	(-)	0.3	1.0	-	-	...
Total	**16.0**	**16.3**	**(7.7)**	29.3	**14.8**	**14.9**	**(6.9)**	42
Developing								
Argentina	0.4	0.1	-	1	0.2	-	-	1
Bolivia	15.8	17.3	(8.2)	20	9.7	13.4	6.2	32
Brazil	50.2	39.1	(18.6)	6	44.2	35.1	16.2	50
Indonesia	31.6	31.7	(15.1)	35	30.2	30.4	14.1	32
Laos	0.3	0.3	(0.2)	1	-	-	-	-
Malaysia	32.0	28.5	(13.5)	40	51.9	49.0	22.7	120
Mexico	0.2	-	-	0.6	4.4	5.0	2.3	7
Myanmar	0.6	0.6	(0.3)	2	0.5	0.3	0.1	1
Namibia	0.5	0.9	(0.4)	1.5	-	-	-	-
Nigeria	1.2	0.2	(0.1)	2	0.3	0.3	0.1	3
Peru	5.1	5.1	(2.4)	8	-	-	-	-
Rwanda	0.7	0.7	(0.3)	2	-	-	-	2
Singapore	-	-	(-)	-	0.6	0.6	0.3	-
S Korea	-	-	(-)	0.1	2.4	2.5	1.2	2
Thailand	14.7	14.6	(6.9)	30	14.6	15.5	7.2	44
Zaire	1.6	1.6	(0.8)	4	-	-	-	1
Zimbabwe	0.8	0.8	(0.4)	2	0.8	0.8	0.4	2
Others	0.1	-	-	0.9	-	2.8	1.3	-
Total	**155.8**	**141.5**	**(67.2)**	210.1	**159.8**	**155.7**	**72.0**	297
Centrally Planned								
China	33.0	35.8	(11.0)	45	28.3	28.0	13.0	45
Czecho-slovakia	0.5	0.3	(0.1)	1	0.1	0.1	..	1
E Germany	2.5	1.8	(0.8)	2	2.5	3.0	1.4	4
Mongolia	1.2	1.2	(0.6)	-	-	-	-	-
USSR	14.0	13.0	(6.2)	20	15.0	14.0	6.5	20
Vietnam	0.8	0.8	(0.4)	1	0.6	0.5	0.2	1
Total	**52.0**	**52.9**	**(25.1)**	69	**46.5**	**45.6**	**21.1**	71
TOTAL	**223.8**	**210.7**		308.4	**221.1**	**216.2**		410

(a) Secondary metal only included for Centrally Planned economies.

TIN

Metal capacity figures refer to primary metal only except for Centrally Planned economies. The capacities for producing metal are very imprecise as smelters can use differing grades of concentrate. The last columns suggest that there is much more capacity available than effectively exists in practice. Much capacity (both mine and metal) closed permanently in 1990-92 or went on standby.

WORLD PRODUCTION OF SECONDARY METAL, 1989-90
('000 tonnes of metal and % of total 1990)

Developed	1989	1990	%	Developing	1989	1990	%	Centrally Planned	1989	1990	%
Australia	0.3	0.3	(1.6)	Argentina	0.1	0.1	(0.5)	USSR	4.0	3.7	(20.1)
Belgium	5.0	6.0	(32.6)	Brazil	0.2	0.2	(1.1)	Others	n.a	n.a	
Canada	0.2	0.2	(1.1)	India	0.2	0.2	(1.1)				
S Africa	0.4	0.3	(1.6)	Thailand	...	...	...				
Spain	0.2	0.2	(1.1)								
UK (a)	7.2	5.9	(32.1)								
USA	0.6	0.5	(2.9)								
Other Europe	0.7	0.8	(4.3)								
Totals	14.6	14.2	(77.2)		0.5	0.5	(2.7)		4.0	3.7	(20.1)
TOTALS		1989	19.1								
		1990	18.4								

(a) Includes production from smelter residues.

This table is incomplete, as full data are not available.

RESERVE/PRODUCTION RATIOS

Static Reserve Life (years):	28
Ratio of identified reserve base to cumulative consumption 1991-2010:	1.8 : 1

CONSUMPTION

	'000 tonnes		% p.a. growth rates		
	1989	1990	1960s	1970s	1980s
European Community	54.5	57.8	-0.1	-1.7	0.5
Japan	33.5	33.8	6.9	1.9	0.9
United States	37.1	37.3	1.3	-1.6	-1.7
Others	60.2	55.0	1.4	2.3	3.4
Total Western World	**185.3**	**183.9**	**1.4**	**-0.2**	**0.8**
Total World	**241.7**	**232.7**	**2.1**	**-0.4**	**0.5**

Includes tin refined from secondary materials, but not use of scrap tin.

END USE PATTERNS 1990(%)

	USA	Japan	UK	W. Germany
Tinplate	26	23	35	} 17
Tinning	4	9	6	
Solder	35	47	10	16
Alloys	7	9	32	} 9
Wrought Tin	7	..	-	
Chemicals & others	28	12	17	58

US and UK - primary and secondary metal; W Germany and Japan - primary only.

Tinplate takes a much greater share(31%) of total world consumption than it does in the major consuming countries shown in the table.

VALUE OF CONTAINED METAL IN ANNUAL PRODUCTION

$1.2 billion (primary refined metal at average 1991 prices).

SUBSTITUTES

Aluminium, tin-free steel, glass, paper, plastics all compete with tin in cans.

Non-metallic materials, copper, aluminium and zinc-coated products are alternatives in roofing and construction applications.

Aluminium alloys, copper-base alloys and plastics can substitute in bronze.

Other chemicals can replace tin compounds for use as fungicides and biocides or polyvinyl chloride stabilisers. In some uses (eg: marine pleasure craft) the use of tin biocides is banned on environmental grounds.

Epoxy resins can be used for solder though not as effectively.

Babbit metal can be replaced by low tin aluminium, copper, or lead bearing alloys and roller or ball bearings.

TECHNICAL POSSIBILITIES

Increasing recovery from slimes in beneficiation stage.

Replacement of tinplate food and drink containers by aluminium can and PET containers.

Increased use of tin oxide coatings.

TIN

PRICES

	1986	1987	1988	1989	1990	1991
LME Standard Grade Cash (a) $/lb	-	-	-	3.69	2.81	2.52
Metal Bulletin free Market Spot $/lb	2.85	3.14	3.26	4.11	-	-
Real Dec. 1991 Prices	3.31	3.55	3.56	4.28	2.81	2.51
Kuala Lumpur/ Penang $M/kg	15.56	16.84	18.49	22.68	16.46	14.94
US$/lb	2.95	3.03	3.20	3.85	2.76	2.48

(a) The LME price was suspended October 24th 1985. The LME resumed quotations in mid 1989 and the 1989 figure refers to the second half.

Until the collapse of the International Tin Agreement (ITA) in October 1985 most tin trade was related to LME or Kuala Lumpur market determined prices, as modified by ITC intervention. Subsequently LME dealings ceased, and prices were fixed by reference to the Kuala Lumpur Exchange or a variety of dealer prices. LME trading resumed in mid 1989.

Up to October 1985 prices were theoretically subject to ITA intervention levels through a system of floor and ceiling prices. The ranges were altered periodically to keep them roughly in line with costs, but the ceiling was often breached. The floor price in the Sixth Agreement was Malaysian $29.16/kg. Prices were supported by export controls and buffer stock purchases. The ITC's prices were set too high in an oversupplied market, in which a substantial and growing percentage of output came from non members. The ITC's buffer stock manager eventually ran out of funds and the Agreement collapsed.

MARKETING ARRANGEMENTS

Production is mainly concentrated in developing countries. Output is largely state controlled in Indonesia, and Nigeria (now a minor producer), and to a lesser extent in Malaysia and Zaire. Brazilian output has expanded very rapidly in the 1980s, largely from small scale 'garimpeiro' mines in the interior. Their output has often been smuggled out of Brazil, and controlling them has proved well nigh impossible.

August 1983 saw the formation of the Association of Tin Producing Countries (ATPC). Its initial aims were the promotion of producer and consumer interests, including research in marketing and the development of new uses for tin. Founder members of Malaysia, Indonesia, Thailand and Bolivia were later joined by Nigeria, Zaire, Australia, and Brazil. China has observer status. The ATPC introduced export quotas after the collapse of the ITC, but these have been more in the nature of mutual comfort than practical constraints. Output has mainly responded to the economic stimuli of weak prices.

The tin market is influenced by sales from the US stockpile (some 7 -10,000 tonnes/year), and by unpredictable Chinese exports (c.15-25000 tonnes/year), and since October 1985 by the unloading of excess stocks built up by the ITC's buffer stock and held by the banks as collateral. These stocks were almost exhausted by 1990-91.

REAL PRICES 1979 to 1991
Tin, Metal Bulletin Free Market/LME

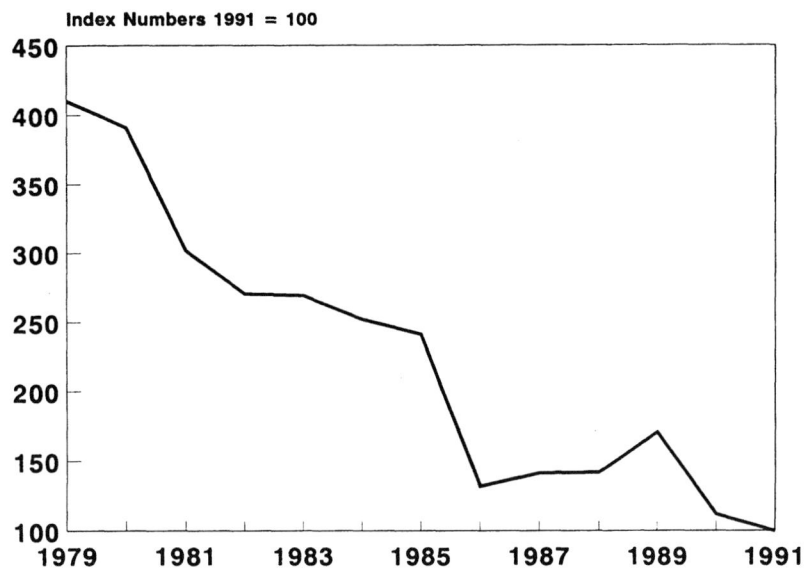

Index Numbers 1991 = 100

WORLD PRODUCTION 1979 to 1991
Tin

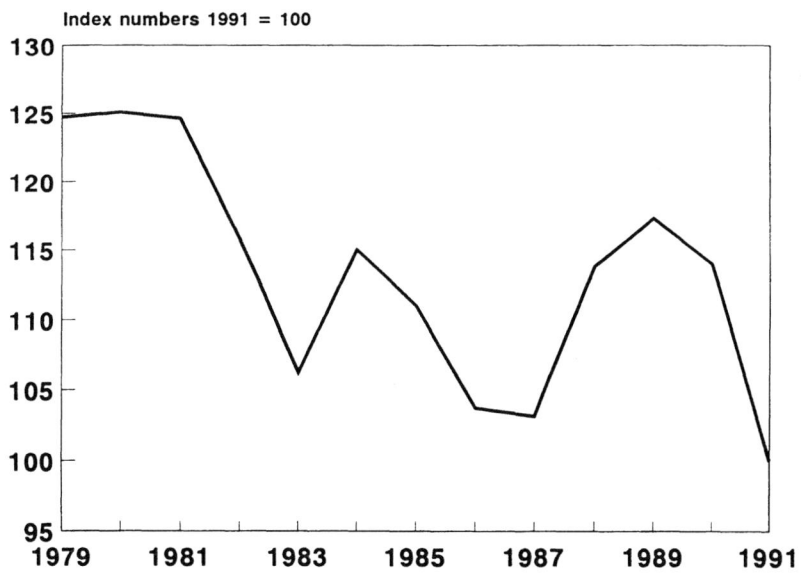

Index numbers 1991 = 100

TIN

SUPPLY AND DEMAND BY MAIN MARKET AREA

	UK 1989	UK 1990	EC(12) 1989	EC(12) 1990	Japan 1989	Japan 1990	USA 1989	USA 1990
Production (tonnes)								
Tin in concentrates	3846	3389	4901	5250	-	-	100	-
Primary metal	3584	6122	10000	12300	808	816	1000	-
Recycled tin metal	7184	5897	7900	7900	-	-	600	500

(a) Includes smelter residues

	UK 1989	UK 1990	EC(12) 1989	EC(12) 1990	Japan 1989	Japan 1990	USA 1989	USA 1990
Net Imports (tonnes)								
Tin in concentrates(gross)	13300	9410	18300	10693	-	-	n.a	n.a
Sn content	5782	3651	n.a	n.a	-	-	216	-
Tin metal	4675	3824	36848	36674	32540	32771	34056	33810

Source of Net Imports (%)

	UK 1989	UK 1990	EC(12) 1989	EC(12) 1990	Japan 1989	Japan 1990	USA 1989	USA 1990
Tin in concentrates								
European Community		24						
Brazil			4					
Bolivia	56	66	40	58				
Burma			5	1				
Nigeria			3	2				
China	12		9	5				
Peru	29	9	24	19				
Zaire			8	11				
Others	3	1	7	4				
Tin Metal								
Australia							4	5
European Community	1	3					2	1
S Africa	6	7	2	1				
USA				1				
Bolivia			1				14	25
Brazil	68	65	35	43			31	19
Chile							1	2
Indonesia			10	13	29	26	15	14
Malaysia	17	15	30	20	40	39	7	12
Nigeria	1							
Singapore			7	9	1	4	1	
Thailand			6	7	18	22	1	
Zimbabwe			1	2				
China		8	5	4	11	7	14	13
Mexico							5	8
Others	6	2	3	1	1	2	5	1

	UK		EC(12)		Japan		USA	
	1989	1990	1989	1990	1989	1990	1989	1990
Net Exports								
(tonnes)								
Tin in concentrates(gross)	1400	998	1800	1000	-	-	-	-
Sn content	1193	967	n.a	n.a	-	-	5457	1201
Tin metal	5371	5661	1000	2000	-	-	391	2095
Consumption								
(tonnes)								
Total Metal	10151	10375	54500	57800	33500	33800	37100	37300
Import Dependence								
Imports as % of								
consumption	100	72	91	83	97	92	91	100
Imports as % of consumption								
and net exports	63	44	87	78	97	97	80	83
Share of World Consumption (%)								
Western world	5	6	29	31	18	18	20	20
Total world	4	4	22	25	14	14	15	16
Consumption Growth (% p.a.)								
1960s	-1.6		-0.1		6.9		1.3	
1970s	-4.9		-1.7		1.9		-1.6	
1980s	0.5		0.5		0.9		-1.7	

TITANIUM

WORLD RESERVES OF ILMENITE
(million tonnes of contained titanium and % of total)

Developed			Developing			Centrally Planned		
Australia	24	(11.8)	Brazil	2	(1.0)	China	30	(14.7)
Canada	27	(13.2)	India	31	(15.1)	USSR	6	(2.9)
Finland	1	(0.5)	Sri Lanka	4	(2.0)			
Norway	32	(15.7)	Mozambique	2	(1.0)			
S Africa	36	(17.6)	Sierra Leone	1	(0.5)			
USA	8	(4.0)						
Totals	**128**	**(62.8)**		**40**	**(19.6)**		**36**	**(17.6)**
Grand Total		**204**						

The world reserve base is estimated at 429 million tonnes. It includes deposits in Egypt, Italy, Madagascar and Malaysia in addition to the above countries. Identified resources total about 1 billion tonnes of contained titanium.

WORLD RESERVES OF RUTILE
(million tonnes of contained titanium and % of total)

Developed			Developing			Centrally Planned		
Australia	5.3	(6.2)	Brazil	66.0	(77.7)	USSR	2.5	(2.9)
S Africa	3.6	(4.2)	India	4.4	(5.2)			
USA	0.3	(0.4)	Sierra Leone	2.0	(2.4)			
			Sri Lanka	0.8	(0.9)			
Totals	**9.2**	**(10.8)**		**73.2**	**(86.2)**		**2.5**	**(2.9)**
Grand Total		**84.9**						

The reserve base totals 165 million tonnes, 52% of which is in Brazil. It also includes deposits in Italy. Identified world resources total 200 million tonnes of contained titanium.

WORLD PRODUCTION OF TITANIUM MINERALS, 1989-90
('000 tonnes of concentrates)

Developed	1989	1990	Developing	1989	1990	Centrally Planned	1989	1990
Ilmenite								
Australia (a)	1714	1619	Brazil	147	150	China	150	150
Norway	930	900	India	160	160	USSR	460	430
USA	260	260	Malaysia	521	502			
			Sierra Leone	62	55			
			Sri Lanka	95	75			
			Thailand (a)	17	11			
Rutile								
Australia	243	226	Brazil	2	3	USSR	10	10
S Africa	60	60	India	5	5			
USA	25	30	Sierra Leone	128	144			
			Sri Lanka	5	5			
Slags								
Canada	1040	760						
S Africa (b)	725	725						

(a) Includes leucoxene.
(b) Ti content of titanium minerals and slag.

TITANIUM

WORLD MINE PRODUCTION, 1989-90, and PRODUCTIVE CAPACITY, 1990
('000 tonnes of contained titanium dioxide and % of total 1990)

	Mine Production		% of Production	Productive Capacity
	1989	1990	1990	1990
Developed				
Australia	1225	1109	(27.1)	1360
Canada	868	837	(20.4)	880
Norway	414	362	(8.8)	450
S Africa	748	742	(18.1)	734
USA	140	140	(3.4)	247
Total	**3355**	**3190**	**(77.8)**	**3671**
Developing				
Brazil	81	63	(1.5)	98
India	100	100	(2.4)	219
Malaysia	224	221	(5.4)	275
Sierra Leone	160	171	(4.2)	170
Sri Lanka	57	42	(1.0)	93
Thailand	10	6	(0.1)	16
Total	**632**	**603**	**(14.7)**	**871**
Centrally Planned				
China	75	75	(1.8)	90
USSR	230	230	(5.6)	260
Total	**305**	**305**	**(7.4)**	**350**
TOTAL	**4292**	**4098**		**4891**

(a) excludes US production of rutile.

88.5% of the total capacity is for ilmenite. Rutile capacity is concentrated in the USA (25 tonnes), USSR (10 tonnes), Sierra Leone (120 tonnes), S Africa (94 tonnes), India (19 tonnes), Sri Lanka (13 tonnes) and Australia (260 tonnes). All figures refer to contained titanium dioxide.

Titanium dioxide contains 60% titanium.

WORLD PRODUCTION OF TITANIUM METAL, 1989-90
(tonnes of sponge and % of total 1990)

Developed	1989	1990	% 1990	Centrally Planned	1989	1990	% 1990
Japan	21.3	25.6	(26.3)	China	2.2	2.2	(2.3)
UK	1.8	1.8	(1.8)	USSR	46.0	43.0	(44.2)
USA	25.2	24.7	(25.4)				
Totals	48.3	52.1	(53.5)		48.2	45.2	(46.5)
TOTALS	1989-	96.5					
	1990-	97.3					

RESERVE/PRODUCTION RATIOS

Static Reserve Life (years):	70 (ilmenite and rutile combined)
Ratio of identified reserve base to cumulative demand 1991-2010: (ilmenite and rutile combined)	7.5 : 1

CONSUMPTION

	'000 tonnes		% p.a. growth rates	
	1989	1990	1970s	1980s
Titanium pigments (at 100% TiO_2)				
European Community	c.830	c.880	1.8	c.1.9
Japan	216	218	4.0	2.8
United States	885(a)	863(a)	1.3	2.5
Titanium sponge				
European Community	23	25	n.a	n.a
Japan	10	11	12.9	1.1
United States	25	23	6.2	-

(a) apparent

TITANIUM

END USE PATTERNS, 1990 (USA) (%)

Ore (Ilmenite and Rutile)

Pigment manufacture (titanium dioxide)	95
Sponge production, welding rod coats and carbides, ceramic and glass formulations	5

Metal

Aircraft and aerospace	75
Chemical processing, power generation, marine and ordnance, medical and other non-aerospace applications	25

VALUE OF ANNUAL PRODUCTION

$1 billion ($TiO_2$ content at average 1991 prices).

SUBSTITUTES

Nickel steels, stainless steels, HSLA steels and some non-ferrous metal alloys can sometimes replace titanium alloys in industrial uses although at the expense of performance or economics.

Tungsten carbide competes with titanium carbide as a cutting surface in machine tools.

Synthetic rutile made from ilmenite can be substituted for natural rutile. Slag competes with ilmenite and rutile.

There are no cost effective substitutes for titanium dioxide pigments.

TECHNICAL POSSIBILITIES

Environmental problems mean that more titanium dioxide plants are likely to use chloride technology in the future. Increasing amounts of synthetic rutile or slag (made from ilmenite) are likely to be used as feed.

Replacement of titanium alloys in aerospace applications by lithium-aluminium alloys or carbon-epoxy composites.

PRICES

	1986	1987	1988	1989	1990	1991
Ore						
Rutile bulk conc 96% TiO_2 Australian fob/fid US$/tonne	404.1	441.0	474.1	527.4	698.4	623.2
Real Dec 1991 prices	469.0	498.3	517.1	548.7	700.1	620.7
Ilmenite Australian bulk conc min 54% TiO_2						
US$/tonne	41.4	52.6	63.8	70.7	73.9	73.6
Real Dec 1991 prices	48.1	59.4	69.6	73.6	74.1	73.4
Sponge						
Spot. Americas $/lb	3.74	3.91	4.82	5.01	4.74	4.0

Source: Metal Bulletin.

Ore prices are mainly fixed on contract. Metal prices are usually quoted by mills. Discounting is common.

MARKETING ARRANGEMENTS

A few large chemical and industrial companies, eg, Du Pont, dominate pigment production, and some have captive ore supplies. They buy feedstock from a similarly small number of suppliers. Production of titanium metal is also controlled by a few firms, mainly in the USA and Japan. The metal/sponge sector was the scene of fierce EEC and US trade actions during the 1980s.

REAL PRICES 1979 to 1991
Titanium, Australian rutile and ilmenite

Index Numbers 1991 = 100

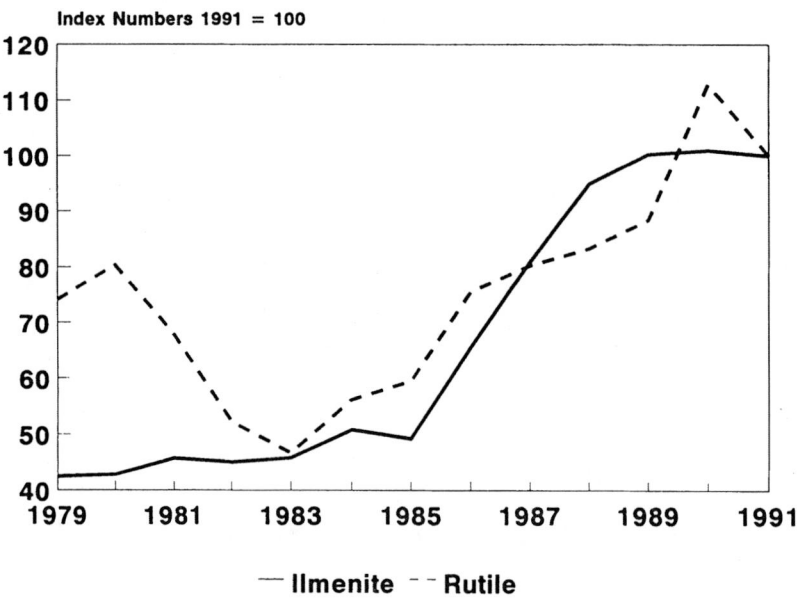

— Ilmenite   - - Rutile

WORLD PRODUCTION 1979 to 1991
Titanium, Concentrates inc. Slag

Index Numbers 1991 = 100

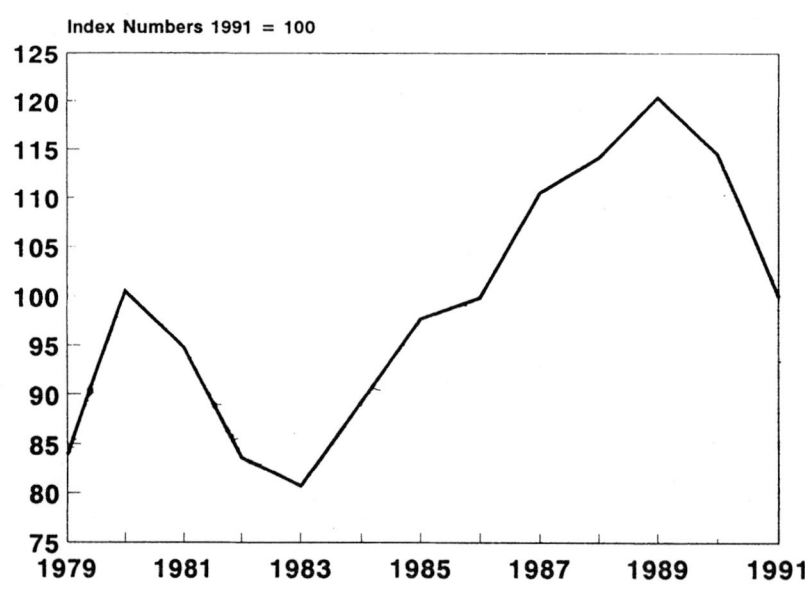

REAL PRICES 1979 to 1991
Titanium Sponge, Americas

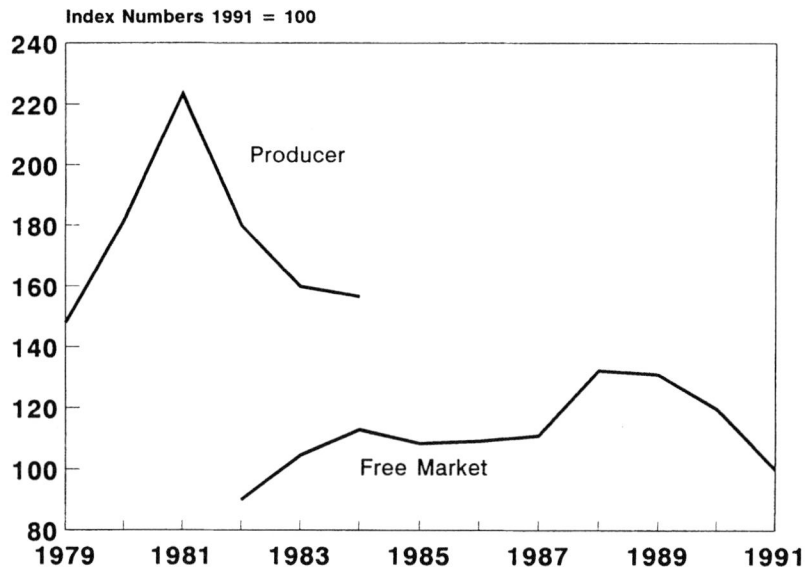

Index Numbers 1991 = 100

Producer

Free Market

WORLD PRODUCTION 1979 to 1991
Titanium Sponge

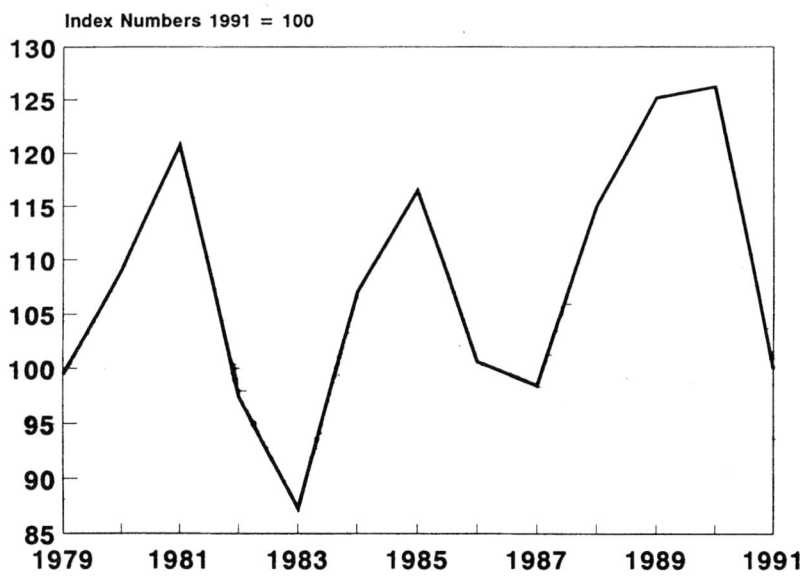

Index Numbers 1991 = 100

TITANIUM

SUPPLY AND DEMAND BY MAIN MARKET AREA

	UK		EC(12)		Japan		USA	
	1989	**1990**	**1989**	**1990**	**1989**	**1990**	**1989**	**1990**
Production ('000 tonnes)								
Ilmenite concentrate	-	-	..	..	-	-	c.260	c.260
Rutile concentrate	-	-	-	-	-	-	c.25	c.30
Titanium slags	-	-	-	-	-	-	-	-
Titanium dioxide pigments	270	270	c.1050	c.1050	c.250	c.250 (TiO$_2$	1007 936	979 913)
Sponge metal	1.8	1.8	1.8	1.8	21.3	25.6	25.2	24.7
Ingots (inc. scrap and imported sponge)	n.a	n.a	n.a	n.a	n.a	n.a	41.3	36.8
Net Imports ('000 tonnes)								
Ilmenite concentrate	380	248	1105	817)	616	758	(459	375
Rutile concentrate	109	127	338	410)			(189	218
Titanium slag and residues	-	-	331	262	n.a	n.a	386.1	373.4
Titanium oxides (inc. pigments)	48.0	39.5	112.6	116.0	71.3	65.2	166.4	147.6
Sponge and metal (inc. wrought)	15.9	17.8	25.8	27.1	1.1	0.8	7.8	5.6
Source of Net Imports (%)								
Ilmenite, rutile and slag								
Australia	57	65	26	27	28	32	41	46
Canada	16		38	31	5	17	16	13
European Community	2	1						
Norway	15	21	18	21				
S Africa			8	9		1	30	32
India		3	3	2	11	17		
Malaysia	2		1	3	39	20		
Morocco								
Sierra Leone	7	9	3	4			9	8
Sri Lanka			2	1	10	6	3	
Thailand					1	2		
USA				2		1		
China					4	4		
Others	1	1	1		2		1	1

	UK 1989	UK 1990	EC(12) 1989	EC(12) 1990	Japan 1989	Japan 1990	USA 1989	USA 1990
Titanium Oxides								
Australia	..		2	9	18	19	1	
Canada		1				27	32	
European Community	64	66			39	32	59	54
Finland	20	18	27	23	2	2	2	2
Japan							3	3
Norway	1	1	8	7			4	4
Singapore					10	7		
United States	12	12	45	44	23	26		
Yugoslavia			3	2				
China	1		6	2	2	2	1	1
Czechoslovakia	1	1	3	2				
South Korea					6	4		
Poland			1	2				
Mexico				5			1	3
Others	1	2	4	4	..	8	2	1
Metal (inc. scrap)								
Austria		1	4	4			2	1
Sweden	2	1	1	1			1	1
Canada							14	17
European Community	18	13			14	5	22	22
Japan	23	26	34	32			42	42
Switzerland								
USA	31	44	34	47	36	39		
China	2	2	1	1			6	7
USSR	20	9	20	10	48	46	7	7
S. Korea						6		
Others	4	4	6	5	2	4	6	3
Net Exports ('000 tonnes)								
Ilmenite concentrate	-	-	2.9	1.6	-)	0.1	(19.8	18.8
Rutile concentrate	n.a	n.a	38.4(a)	19.9(a)	-)		(	
Oxides & titanium dioxide pigments etc.	168.6	144.7	331.7(b)	270.8(b)	86.4	91.0	212.2	270.4
Sponge and metal (incl. wrought)	4.6	4.7	4.8	3.6	13.6	13.2	13.3	13.7

(a) Excl. UK
(b) Excl. Italy

TITANIUM

	UK		EC(12)		Japan		USA	
	1989	1990	1989	1990	1989	1990	1989	1990
Consumption ('000 tonnes)								
Ilmenite	n.a	n.a	n.a	n.a	n.a	n.a	659.6	688.9
						(TiO$_2$	418.7	446.2)
Rutile (natural & synthetic)	n.a	n.a	n.a	n.a	n.a	n.a	366.1	369.5
						(TiO$_2$	346.0	347.1)
Slag	n.a	n.a	n.a	n.a	n.a	n.a	414.8	390.5
						(TiO$_2$	335.4	313.6)
Titanium dioxide pigments (TiO$_2$ content)	145.6	127.5	c.830	c.880	216	218	947.3 (TiO$_2$ 884.6	925.4 863.2
Sponge metal	c.3	c.4	c.23	c.25	10.3	11.0	24.9	23.2
Import Dependence(all forms) Imports as % of consumption	100	100	100	100	100	100	85	84
Imports as % of consumption and net exports	100	100	100	100	100	100	69	65
Share of World Consumption (%) Total world:								
Sponge and metal	3	4	24	26	11	11	26	24
Pigments (approx.)	5	4	27	30	7	8	31	32
Consumption Growth (% p.a.) 1970s								
Metal	n.a		n.a		12.9		6.2	
Pigments etc.	1.0		1.8		4.0		1.3	
1980s								
Metal	n.a		n.a		1.1		-	
Pigments etc.	0.6		c.1.9		2.8		2.5	

TUNGSTEN

WORLD RESERVES
('000 tonnes of metal and % of total)

Developed			Developing			Centrally Planned		
Australia	56	(2.4)	Bolivia	58	(2.5)	China	1050	(44.7)
Austria	10	(0.4)	Brazil	20	(0.9)	Czecho-		
Canada	260	(11.1)	Myannar	15	(0.6)	slovakia	5	(0.2)
France	20	(0.8)	S Korea	58	(2.5)	N Korea	80	(3.4)
Portugal	26	(1.1)	Malaysia	17	(0.7)	USSR	280	(11.9)
Turkey	65	(2.8)	Mexico	8	(0.3)			
UK	70	(3.0)	Thailand	30	(1.3)			
USA	150	(6.4)	Zimbabwe	5	(0.2)			
Others	20	(0.8)	Others	47	(2.0)			
Totals	**677**	**(28.8)**		**258**	**(11.0)**		**1415**	**(60.2)**
Grand Total		**2350**						

The reserve base is 3.44 million tonnes, distributed broadly as above.

TUNGSTEN

WORLD MINE PRODUCTION, 1989-90, and PRODUCTIVE CAPACITY, 1990
('000 tonnes of contained tungsten and % of total 1990)

	Mine Production		% of Production	Productive Capacity
	1989	1990	1990	1990
Developed				
Australia	1.3	1.09	(2.6)	2.25
Austria	1.13	1.41	(3.3)	1.30
Canada	-	-	(-)	3.00
Japan	0.29	0.25	(0.6)	0.30
New Zealand	0.01	-	(-)	0.01
Portugal	1.38	1.41	(3.3)	1.40
Spain	0.06	0.03	(0.1)	0.10
Sweden	0.08	-	(-)	0.35
Turkey	0.05	0.05	(0.1)	0.20
USA	0.4	0.45	(1.0)	3.70
Total	**4.75**	**4.69**	**(10.9)**	**12.61**
Developing				
Argentina	...	0.01	(...)	0.04
Bolivia	1.12	0.82	(1.9)	1.30
Brazil	0.79	0.42	(1.0)	0.50
Myanmar	0.46	0.45	(1.0)	0.50
India	0.01	0.01	(...)	0.02
S Korea	1.56	1.25	(2.9)	1.50
Mexico	0.17	0.18	(0.4)	0.30
Peru	0.27	0.39	(0.9)	1.20
Rwanda	0.10	0.16	(0.4)	0.15
Thailand	0.65	0.29	(0.7)	0.50
Uganda	..	..	(..)	0.01
Zaire	0.02	0.02	(...)	0.03
Zimbabwe	..	..	(..)	0.03
Total	**5.15**	**4.00**	**(9.3)**	**6.09**
Centrally Planned				
China	28.0	25.0	(58.1)	28.0
Czechoslovakia	0.05	0.05	(0.1)	0.05
Mongolia	2.0	1.5	(3.5)	0.50
N Korea	0.3	0.5	(1.2)	1.00
Vietnam		0.3		n.a
USSR	7.0	7.0	(16.3)	9.20
Total	**37.35**	**34.35**	**(79.8)**	**38.75**
TOTALS	**47.25**	**43.04**		**57.45**

RESERVE/PRODUCTION RATIOS

Static Reserve Life (years):	55
Ratio of identified reserve base to cumulative demand 1991-2010:	5.7 : 1

CONSUMPTION

Tungsten Concentrates

	tonnes(Tungsten Content)		%p.a. growth rates	
	1989	**1990**	**1970s**	**1980s**
European Community	3278	1299	-6.0	-12.1
Japan	1538	1440	-4.9	-6.9
United States	7725	5878	1.64	-4.5
Other Countries	5659	5320	4.1	-2.9
Total Western World	**18200**	**13937**	**1.1**	**-5.3**
Eastern Countries	29800	29000	n.a	1.7
Total World	**48000**	**42937**	**0.6**	**-1.4**

Source: UNCTAD Committee on Tungsten.

These statistics show the immediate consumption of tungsten concentrates, but not necessarily the final destination of the subsequent products. Trade is increasingly in tungsten products, and especially ammonium paratungstate (APT). China has become the world's dominant supplier.

All Forms
'000 tonnes tungsten content

	1988	1989	1990
Total Western World	29	30	27
Other Countries	19	20	20
Total World	**48**	**50**	**47**

Source: E & MJ March 1992.

END USE PATTERNS, 1990 (%)

United States		**Japan**	
Metal working and construction machinery	68		
Transport & Electrical	12		
Lamps and lighting	12		
Chemicals	5		
Other	3		
		Tungsten Metal	
Cemented carbides	58	Cemented carbides	84
Tool steels	4	Wire, sheets, rods	1
Tungsten metal	26	Contacts	3
Superalloys and miscellaneous	12	Special steel	1
		Others	1

TUNGSTEN

VALUE OF CONTAINED METAL IN ANNUAL PRODUCTION
$244 million (contained WO_3 at average 1991 prices).

SUBSTITUTES

Titanium, tantalum and niobium carbides can be used in some wear-resisting applications.

Molybdenum tool steels and tungsten tool steels are interchangeable.

In some cutting tool applications, bulk ceramics are alternatives.

TECHNICAL POSSIBILITIES

Further development of new metal-shaping methods, eg: laser. Development of new cutting tool materials.

Increased use in ceramics and in catalysts.

Increased use of tungsten scrap.

Increased use of coatings on cemented carbide cutting tools to prolong tool life.

Use of tungsten compounds in light-sensitive applications.

PRICES

	1986	1987	1988	1989	1990	1991
Ore min 65% WO_3 cif Europe						
$/mtu WO_3	47.4	48.9	56.0	56.7	46.3	56.7
Real Dec 1991 prices	55.0	55.3	61.1	59.0	46.3	56.5
International Tungsten Indicator						
US$/mtu	55.4	51.5	59.0	60.9	54.1	-
Real Dec 1991 prices	64.2	58.2	64.4	63.4	54.4	-

Source: Metal Bulletin

Prices are traditionally highly volatile. Consumers buy direct from producers or through traders, often by reference to Metal Bulletin price quotes which usually represent small spot lots or samples of purchases. An alternative pricing basis was the International Tungsten Indicator which moved in a very similar fashion. It was put 'on hold' from December 1990. The Metal Bulletin reorganised its price quotations in 1991, tightening the specifications and introducing new products.

MARKETING ARRANGEMENTS

The market is dominated by exports from China, especially of ammonium paratungstate and most Western primary mines have closed or are on standby. The USA concluded an orderly Marketing Agreement with China in 1987. It expired in September 1991. An antidumping case in the US in 1991 resulted, however, in the imposition of 151% antidumping duties on Chinese material. The European Community reached a negotiated settlement with China over Chinese exports which imposed some duties and raised prices. China suspended exports of tungsten concentrates in early 1991, but trade had anyway largely switched to products.

Sales from the US stockpile were suspended between October 1985 and May 1986 and then resumed on only a moderate basis until 1989, when 466 tonnes were sold. The total inventory now falls short of the redefined goal.

Major producers meet under the auspices of the International Tungsten Industry Association which includes major customers and traders. The UNCTAD Tungsten Committee collected data and attempted unsuccessfully, so far, to reach agreement on market stabilisation measures.

REAL PRICES 1979 to 1991
Tungsten, Ore cif Europe

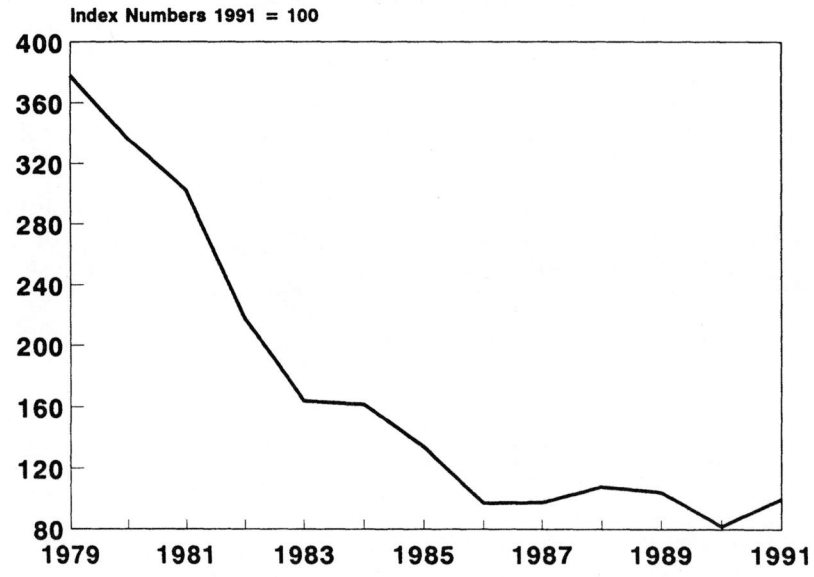

Index Numbers 1991 = 100

WORLD PRODUCTION 1979 to 1991
Tungsten

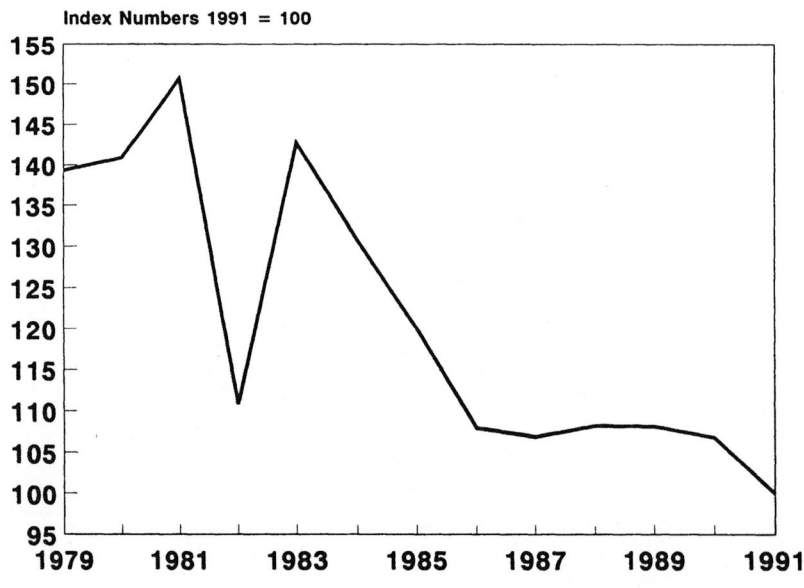

Index Numbers 1991 = 100

SUPPLY AND DEMAND BY MAIN MARKET AREA

	UK		EC(12)		Japan		USA	
	1989	1990	1989	1990	1989	1990	1989	1990
Production (tonnes)								
Mine production (W content)	35	56	1464	1438	286	254	400	450
Ammonium paratungstate (W content)	n.a	n.a	n.a	n.a	n.a	n.a	7831	6142
Shipment of excess material from US govt stockpile (W content)							466	-
Net Imports (tonnes)								
Ores Concentrates (gross wt)	85	111	6970	1338	2316	2589	15550	11372
Concentrates (W content)	44	45	2935	1092	1389	1554	7837	6126
Ferro tungsten (gross wt)	78	190	2358	2209	743	835	552	671
Ammonium paratungstate (W content)	1426	787	5055	2670	3787	3342	591	
Tungsten metal & powders scrap	631	735	2340	2268	378	348	2212	1708
Tungsten carbide (gross wt)	473	294	1981(a)	1140(a)	n.a	n.a	703	736
All Forms (W content)	2087	1762	15612(b)	13114(b)	5109	4954	11801	9982

(a) Excl. Denmark.
(b) Incl. intra Community trade.

Source of Net Imports (%)

Ores and Concentrates

	UK		EC(12)		Japan		USA	
	1989	1990	1989	1990	1989	1990	1989	1990
Australia			2		14	22	1	
Canada								
European Community	16				47	49	4	3
Norway								
Sweden			10					
Hong Kong							3	2
Mauritania								
China	20	5	74	32	27	14	72	63
Bolivia	22		2		4	6	8	15
Myanmar								
Turkey							1	
Mexico							2	3
Peru					9	5	8	10
Rwanda					1	1		2
S Korea					1			
Thailand		4		1	3	1		
Vietnam		2						
Others and undefined	42	95	6	68	2			2

TUNGSTEN

	UK 1989	UK 1990	EC(12) 1989	EC(12) 1990	Japan 1989	Japan 1990	USA 1989	USA 1990
Net Exports (tonnes)								
Ores Concentrates (gross wt)	41	24	1330	69299	-	-	125	72
Concentrates (W content)	13	17	1328	1476	-	-	203	139
Metal and powder (W content)	748	652	1990(a)	1860(a)	144	163	2066	1831
Tungsten carbide (gross wt)	463	461	325 (a)	285 (a)	-	-	1360	1074
Ferro tungsten (gross wt)	21	4	27	11	-	-	42	31
All Forms (W content)	1197	1226	12062(b)	7741(b)	148	166	3388	3183

(a) Excl. W Germany.
(b) Incl. intra Community Trade.

	UK 1989	UK 1990	EC(12) 1989	EC(12) 1990	Japan 1989	Japan 1990	USA 1989	USA 1990
Consumption (tonnes)								
Concentrates (W content)	50	50	3278	1299	1538	1440	7725	5878
Apparent Total (W content)	909	558	7060	7056	4822	5287	10474	8287

(a) Concentrates, scrap and metal.

	UK 1989	UK 1990	EC(12) 1989	EC(12) 1990	Japan 1989	Japan 1990	USA 1989	USA 1990
Import Dependence (all forms)(a)								
Imports as % of consumption	100	100	100	100	100	100	100	100
Imports as % of consumption and net exports	99	99	82	89	100	91	85	87

	UK 1989	UK 1990	EC(12) 1989	EC(12) 1990	Japan 1989	Japan 1990	USA 1989	USA 1990
Share of World Consumption (%)								
Concentrates								
Western World	..	..	18	9	8	10	42	42
Total World	..	..	7	3	3	12	16	19
All Forms								
Western World	3	2	23	26	16	20	35	31
Total World	2	1	14	15	10	11	21	18

	UK	EC(12)	Japan	USA
Consumption Growth (% p.a.)				
Concentrates				
1970s	-8.3	-6	-4.9	1.6
1980s	-28.6	-12.1	-6.9	-4.5
All Forms				
1970s	n.a	n.a	n.a	n.a
1980s	n.a	n.a	6.7	-1.8

URANIUM

WORLD REASONABLY ASSURED RESOURCES
('000 tonnes uranium and % of total)

Developed			Developing			Centrally Planned		
Australia	538	(18,8)	Algeria	26.0	(0.9)	Bulgaria	5	(0.2)
Canada	235	(8.2)	Argentina	11.6	(0.4)	Czechoslovakia	40	(1.4)
Denmark	27	(0.8)	Brazil	163.0	(5.7)	Hungary	2	(0.1)
Finland	1.5	(0.1)	Central African			Romania	18	(0.6)
France	58.9	(2.1)	Republic	16.0	(0.6)	USSR	435	(15.2)
W Germany	4.8	(0.2)	Gabon	17.6	(0.6)	China	50	(1.7)
Greece	0.3	(..)	India	47.2	(1.6)			
Italy	4.8	(0.2)	S Korea	11.8	(0.4)			
Japan	6.6	(0.2)	Mexico	7.7	(0.3)			
Portugal	8.7	(0.3)	Namibia	106.9	(3.7)			
S Africa	418.5	(14.6)	Niger	175.9	(6.1)			
Spain	35.0	(1.2)	Peru	1.8	(0.1)			
Sweden (a)	4.0	(0.1)	Somalia	6.6	(0.2)			
Turkey	3.9	(0.1)	Zaire	1.8	(0.1)			
USA	376.0	(13.1)						
Totals	**1723.0**	**(60.1)**		**593.9**	**(20.7)**		**550**	**(19.2)**
Grand Total		**2866.9**						

Source: OECD/IAEA Uranium Resources, Production & Demand. Uranium Institute

(a)Of which 35 is legally inaccessible on environmental grounds.

After allowing for processing losses not incorporated in these estimates, the total is about 2750.

The table includes estimates of reasonably assured resources available at 1st January 1989. Some 70% of the total is available at an estimated forward cost (not price) of $80/kg U or less, and the balance at a forward cost of $80 to 130 kg/U. Estimated additional resources in the Western world available at a forward cost under $130/kg of U, amount to some 1.16 million tonnes of U, after allowing for losses. There is a further 0.6 million tonnes in Centrally Planned Economies. The estimates for Centrally Planned Economies are on different definitions from those of the Western World. They refer to 'proven' resources, recoverable at costs up to $80/kg U, and may cover rather more than the Western 'reasonably assured resources'.

URANIUM

WORLD MINE PRODUCTION, 1989-90
(tonnes U and % of total 1990)

Developed	1989	1990	% 1990	Developing	1989	1990	% 1990	Centrally Planned	1989	1990	% 1990
Australia	3710	3535	(7.5)	Argentina	51	8	(...)	Bulgaria	600	500	(1.1)
Canada	11356	8721	(18.5)	Brazil	30	4	(...)	China	1300	1300	(2.8)
France	3290	2816	(6.0)	Gabon	170	710	(1.5)	Czecho-			
								slovakia	2600	2400	(5.1)
W Germany	30	-	-	India	230	230	(0.5)	E.Germany	3830	2972	(6.3)
Portugal	145	111	(0.2)	Namibia	3077	3211	(6.8)	Hungary	530	520	(1.1)
S Africa	2957	2478	(5.2)	Niger	2965	2830	(6.0)	Romania	200	200	(0.4)
Spain	226	216	(0.5)					USSR	11000	11000	(23.3)
USA	5151	3387	(7.2)								
Yugoslavia	85	70	(0.1)								
Totals	26950	21334	(45.2)		7223	6993	(14.8)		20000	18892	(40.0)
TOTALS	1989-	54233									
	1990-	47219									

Source: Uranium Institute

Note:There is presently a limited amount of reprocessing of spent reactor fuel which supplements mine production. In addition Belgium produced 40 tonnes U in 1989 and 1990 from imported phosphates.

RESERVE/PRODUCTION RATIOS

Static Reserve Life (years): 58
Ratio of resource base to cumulative demand 1991-2010 (includes reasonably assured and estimated additional resources): 3.7 : 1

CONSUMPTION

The table shows average world reactor requirements based on estimated reactor usage. Stockpiling is ignored.

	'000 tonnes(U)		% p.a. growth rates	
	1989	1990	1970s	1980s
European Community	15.0	14.9	13.9	4.5
Japan	6.1	5.8	27.6	3.8
United States	14.4	15.3	16.6	7.6
Others	7.1	7.1	36.0	4.1
Total Western World	**42.6**	**43.1**	**18.4**	**5.3**
Total World	**n.a**	**50.2**	**n.a**	**n.a**

Source: Uranium Institute: Uranium in the New World Market 1990-2010 & 1992 Update.

The future growth in demand depends on the continuation of programmes for building nuclear reactors. Because of delays to reactor construction and stockpiling, purchases of uranium substantially exceeded reactor usage up to 1983. Public acceptability of nuclear power and of nuclear waste disposal is vital for future growth of demand for uranium. This remains in some doubt.

END USE PATTERNS

The only uses for natural uranium are for military purposes, for research, and for civil nuclear power. There is no published breakdown of demand, although nearly all recent output has gone into the latter.

VALUE OF ANNUAL PRODUCTION

$2.2 billion at an average value of $18/lb U_3O_8 (a weighted average of spot and contract prices) in 1991.

SUBSTITUTES

Nuclear power directly competes with other forms of electricity generation. Once a nuclear station is built, however, there is no substitute for uranium based fuel beyond various fuel management schemes that raise the productivity of the installed fuel. Reprocessing of spent fuel allows some limited substitution for mined output in the longer term, but it is constrained in the short term by a lack of reprocessing facilities and waste storage.

TECHNICAL POSSIBILITIES

Improvements in the technology of existing reactor types will allow savings of up to 15% in uranium requirements. Longer term, the fast breeder reactor, now at the prototype stage, would allow a substantial (60 fold) reduction in uranium usage per unit of power. Its development has been delayed and it will probably not be a commercial proposition until well into the 21st century. Work on different types of reactor from the light water reactor has been inhibited by developments in the energy and uranium markets but several processes have been developed for processing and enrichment.

PRICES

	1986	1987	1988	1989	1990	1991
Nuexco exchange value $/lb U_3O_8	17.0	16.8	14.9	10.2	9.7	8.77
Real Dec 1991 prices	19.1	18.4	15.8	10.6	9.74	8.74

Average Contract
Prices (Source Nukem & Euratom)
$/lb U_3O_8

	1986	1987	1988	1989	1990	1991
USA Domestic Suppliers	30.01	27.37	26.15	19.56	15.70	n.a
Imports	20.07	19.14	19.03	16.75	12.55	n.a
Europe	31.0	32.5	31.82	29.35	29.39	26.09

REAL PRICES 1979 to 1991
Uranium, Nuexco Exchange Value

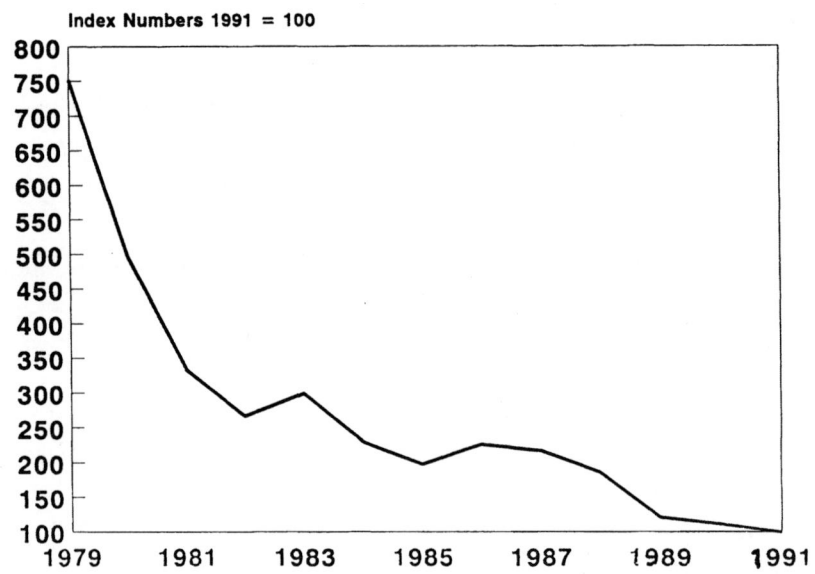

WESTERN WORLD PRODUCTION 1979 to 1991
Uranium

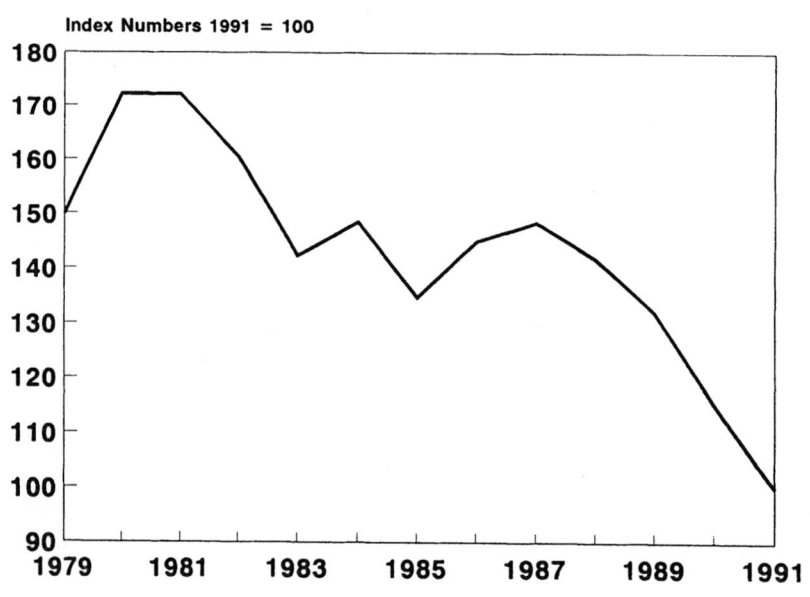

MARKETING ARRANGEMENTS

Uranium is a minor but very important element of the cost of nuclear power. Roughly 75% of sales are under long term contracts directly negotiated between mines and utilities. The intermediate processing from uranium concentrate to fuel rods is carried out under tolling arrangements. Because of their sensitive nature, uranium exports are heavily controlled by host governments in nearly all respects. There is a spot market which has gradually increased in importance and which is organised mainly by uranium brokers in which utilities can dispose of excess supplies. Spot prices are only one influence on prices fixed under long term contracts. Production costs are also important. There is a limited number of mines, which are controlled by a mixture of government agencies, utilities, and mining companies. Most producers have large interests outside uranium mining. The spot market has been increasingly influenced by ever more sophisticated lending between those with excess stocks and those with immediate needs, and by rising exports from Eastern Europe, China and the USSR, not just of natural uranium but also of converted or enriched material. The US imposed provisional antidumping duties on USSR materials in mid 1992. Euratom introduced informal quota guidelines. The growing low priced imports, however, forced the closure of many Western mines in 1989-92.

SUPPLY AND DEMAND BY MAIN MARKET AREA

Data on foreign trade compatible with production and consumption statistics are not published. On the basis of mine production and reactor usage Japan and the United Kingdom are completely reliant on imports or accumulated inventories. Production in the European Community equals about 20% of its reactor usage, whilst in the United States average production greatly exceeded final consumption in nuclear reactors until recently. Demand is met now partly from stocks and from imports. The processing of natural uranium (conversion, enrichment and fuel fabrication) is a constraint on commercial freedom. Supplying countries also impose various restrictions on trade in, and the uses of their exports.

VANADIUM

WORLD RESERVES
('000 tonnes of contained vanadium and % of total)

Developed			Developing		Centrally Planned		
Australia	32	(0.7)	Chile	n.a	China	600	(14.2)
S Africa	862	(20.2)	India	n.a	USSR	2631	(61.7)
USA	135	(3.2)	Peru	n.a			
			Venezuela	n.a			
Totals	1029	(24.1)		n.a.		3238	(75.9)
Grand Total		4267					

The world's reserve base is some 16.6 million tonnes, with over 70% in S Africa and USSR. Norway, Canada, New Zealand, Burundi, Namibia, Mozambique, Zambia, Mexico and South Korea make up 460,000 tonnes of the reserve base.

The world's identified resources amount to 63 million tonnes. Most of the resources are in titaniferous-magnetites from which vanadium would be produced as a by-product of iron, or they are in crude petroleum and tar sands. In all these cases extraction depends on economic recovery of the main product.

WORLD MINE PRODUCTION, 1989-90, and PRODUCTIVE CAPACITY,1990
(tonnes of contained vanadium and % of total)

	Mine Production		% of Production	Productive Capacity
	1989	1990	1990	1990
Developed				
Australia	-	-	-	1500
S Africa	16500	17000	(53.8)	27200
USA	c.1200	c.500	(1.6)	1800
Total	**17700**	**17500**	**(55.4)**	**30500**
Developing				
Chile	-	-	-	2300
Total				**4800**
Centrally Planned				
China	4500	4500	(14.2)	8200
USSR	9600	9600	(30.4)	9600
Total	**14100**	**14100**	**(44.6)**	**17700**
TOTAL	**31800**	**31600**		**48200**

Includes capacity of plants on standby as well as those operating. Does not include V_2O_5 in iron slags and petroleum refinery residues. Vanadium pentoxide (V_2O_5) contains 56% vanadium.

Japan recovered 560 tonnes of vanadium in both 1989 and 1990, and the USA a further 2308 tonnes in 1990 (2389 tonnes in 1989), from petroleum residues, ashes and spent catalysts. The countries of origin of the raw material are unknown. W Germany, USSR and several European countries also recover vanadium from petroleum residues but insufficient data are available to allow reliable estimates to be made of either output or capacities. Canada began producing from oil sand wastes in late 1990 for commercial sales in 1991.

RESERVE/PRODUCTION RATIOS

Static reserve life (years):   135
Ratio of identified reserve base to
cumulative demand 1991-2010:   23 : 1(excluding petroleum residues)

CONSUMPTION

	tonnes		% p.a. growth rates	
	1989	1990	1970s	1980s
European Community	c.9000	c.9000	n.a probably fell	n.a
Japan	c.3200	c.3300	8.7	n.a.
United States (a)	4646	4098	-0.3	-6.1

(a) Reported consumption.

END USE PATTERNS, 1990 (USA) (%)

HSLA steels	27	Machinery & Tools	40
Other steels	55	Building &	
		Heavy Construction	22
Non-ferrous alloys	12	Transport	22
Others, including cast irons,			
chemicals and catalysts	6	Others	16

VALUE OF ANNUAL PRODUCTION

$400 million (contained metal at average 1991 prices including recovery from residues etc).

SUBSTITUTES

Niobium, molybdenum, titanium, chromium, manganese and tungsten can sometimes substitute for vanadium in steel although at higher cost or with lower performance. Heat-treated carbon steels can replace vanadium steels in some applications.

Platinum can be used in some catalytic processes but at higher cost.

TECHNICAL POSSIBILITIES

Possible recovery from low grade dolomitic shales and sandstones and extraction from tar sands and oil shales.

Potential new chemical applications.

Extended use of HSLA steels and vanadium superalloys.

VANADIUM

PRICES

	1986	1987	1988	1989	1990	1991
Pentoxide/Metallurgical(a)						
Highveld Fused Europe $/lb V_2O_5	2.59	2.70	3.27	4.87	3.35	2.99
Real Dec 1991 prices	3.01	3.05	3.57	5.06	3.36	2.98
Ferrovanadium						
US Producer 80% V $/lb	6.24	6.54	8.36	11.74	8.49	7.95
Real Dec 1991 prices	7.25	7.39	9.12	12.21	8.53	7.92

(a) Source: Metal Bulletin.

The dominant method of pricing is by producers.

MARKETING ARRANGEMENTS

Vanadium is mainly produced as by-product or co-product of other metals, such as iron and uranium. S Africa, especially Highveld Steel and Vanadium Corp, and USSR are the world's major producers, with China becoming increasingly important. Highveld is the market leader. The trend is towards vertical integration through ferrovanadium production facilities. US producers of ferrovanadium are facing increasing competitive pressures from lower cost imports.

The entry of new South African producers in 1990-91, the start up of the Canadian operation and the threat of Australian production cast shadows over the market and contributed to the 1990-91 collapse of prices.

REAL PRICES 1979 to 1991
Vanadium, Highveld Pentoxide in Europe

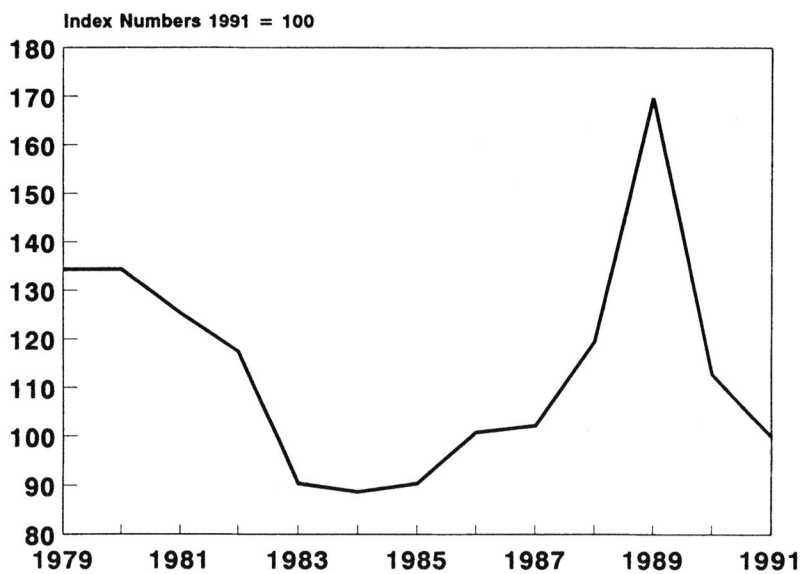

Index Numbers 1991 = 100

WORLD PRODUCTION 1979 to 1991
Vanadium

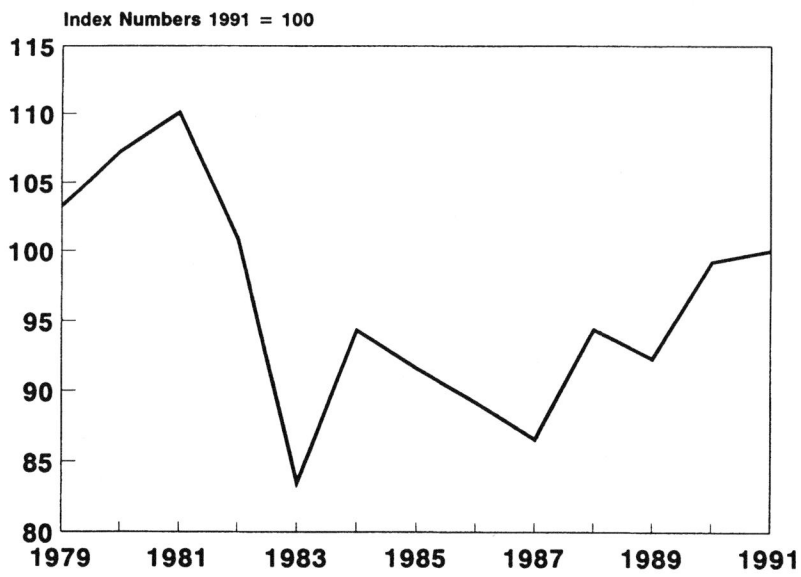

Index Numbers 1991 = 100

VANADIUM

SUPPLY AND DEMAND BY MAIN MARKET AREA

	UK		EC(12)		Japan		USA	
	1989	1990	1989	1990	1989	1990	1989	1990
Production								
(tonnes)								
Ores and concentrates:								
Mine (V content)	-	-	-	-	-	-	c.1200	c.500
Oil Residues &								
spent catalysts								
(V content)	-	-	-	-	560	560	2389	2308
Ferrovanadium	-	-	n.a	n.a	3127	c.3200	n.a	n.a
Net Imports								
(tonnes)								
Ores, slags and								
residues (gross)	31	26	13301	442	728(a)	127(a)	4419	5253
(V content)							3157	3846
Vanadium oxides (gross)	256	226	7362	6503	4313	4796	448	1004
(V content)							250	680
Ferrovanadium (gross)	824	631	1694	2640	238	431	646	323
(V content)							487	305
Vanadium metal(incl.wrought)	312	257	42	99	154	104	256	115

(a) Ores & concentrates included with tantalum.

Source of Net Imports(%)

	UK		EC(12)		Japan		USA	
	1989	1990	1989	1990	1989	1990	1989	1990
Vanadium Oxides								
Austria	41		2	..				
European Community	3	6			1		12	9
Namibia		41		2				
S Africa	54		37	35	93	79	86	91
Switzerland								
United States		32	12	21	1	7		
China	2	21	49	40	5	14	-	
Others and unspecified		-		2				2
Ferrovanadium								
Austria	46	41	90	85	19	55	17	36
Canada							26	39
European Community	52	59			65	39	43	18
USSR				9				
United States			2					
Brazil					16	6	2	
China							1	5
Japan							9	
Others	2		8	6			2	2

	UK		EC(12)		Japan		USA	
	1989	1990	1989	1990	1989	1990	1989	1990
Net Exports (tonnes) Ores, slags & Residues (gross)	94	20	1585	3983	n.a	n.a	4962	3015
Vanadium oxides(gross) (gross)	18	31	274	98	163	46	3150	2570
Ferrovanadium (gross)	60	231	1179(a)	1488(a)	188	145	512	334
Vanadium metal (incl.wrought)	44	22	154	116	3.5	5.6	42	4

(a) Excludes Belgium-Luxembourg.

	UK		EC(12)		Japan		USA	
Consumption (tonnes)	720	c.520 (V content)	c.9000	c.9000 (V content)	c.3200	c.3200 (V content)	4646	4098 (reported)
Import Dependence Imports as % of consumption	100	100	100	100	100	100	80	100
Imports as % of consumption and net exports	100	100	100	100	100	100	46	61
Share of World Consumption (%) Total world (approx)	2	2	c. 26	c. 25	c.9	c.9	13	12
Consumption Growth (% p.a.) 1970s	-3	(iron & steel industry)	n/a	(probably fell)	8.7		-0.3	
1980s	n.a		n.a		n.a		-6.1	

VERMICULITE

WORLD RESERVES

Western world reserves are estimated at 45 million tonnes, 50% in the United States, 40% in South Africa and 7% in Brazil and other countries. Substantial deposits exist in China, the USSR and other centrally planned economies, but their combined material has generally inferior exfoliation characteristics.

The reserve base is 180 million tonnes and total world resources are approximately 550 million tonnes.

WORLD MINE PRODUCTION, 1989-1990
('000 tonnes and % of total 1990)

Developed	1989	1990	% 1990	Developing	1989	1990	% 1990	Centrally Planned	1989	1990	% 1990
Australia	1.2	1.2	(0.2)	Argentina	0.6	20	(3.6)	USSR	25	25	(4.5)
Japan	15.0	15.0	(2.7)	Brazil	28.0	28	(5.9)	China	25	35	(6.3)
S Africa	195.5	217.7	(39.1)	Egypt	0.1	0.1	(..)				
USA	249.5	208.6	(37.4)	India	2.8	3.0	(0.5)				
				Kenya	2.4	2.5	(0.5)				
				Mexico	0.3	0.3	(0.1)				
				Zimbabwe	1.5	1.0	(0.2)				
Totals	461.2	442.5	(79.4)		29.7	54.9	(9.8)		50	60	(10.8)
Totals	1989-	540.9									
	1990-	557.4									

Percentages shown are for Western world only.
In addition, Malawi and Tanzania are believed to produce vermiculite.

PRODUCTIVE CAPACITY, 1990

Western world productive capacity totals 630,000 tonnes of which 320,000 tonnes is located in the USA and 235,000 tonnes in S Africa. Chinese capacity approaches 100,000 tonnes, but half is fairly primitive. USSR capacity is near 25,000 tonnes.

RESERVE/PRODUCTION RATIOS

Static Reserve Life (years):	81
Ratio of identified reserve base to cumulative demand 1991-2010:	13.6 : 1

CONSUMPTION

	'000 tonnes		% p.a. growth rates	
	1989	1990	1970s	1980s
European Community	c.2000	c.2000	n.a	n.a
Japan	c. 27	c. 27	n.a	n.a
United States (apparent consumption)	281	236	1.9	-2.4

END USE PATTERNS, 1990 (USA) (%)

Exfoliated vermiculite:
Agriculture	40
Insulation	26
Lightweight concrete aggregate	19
Plaster and cement premises	13
Other	5

VALUE OF ANNUAL PRODUCTION

$ 50 million (at average 1991 prices).

SUBSTITUTES

In lightweight concrete and plaster, expanded perlite, expanded clay shale, or slate substitute. Fibreglass, foam and slag wool are amongst the alternatives for insulation, and a variety of substitutes is available in agricultural applications.

TECHNICAL POSSIBILITIES

Newly designed furnaces with higher throughput and improved thermal efficiency are becoming available.

The development of new composite materials is increasing the utilisation of vermiculite.

Development of new uses for finer sized exfoliated vermiculite.

The need for a safer alternative to asbestos has prompted much research on vermiculite.

VERMICULITE

PRICES

	1986	1987	1988	1989	1990	1991
Average value of concentrate fob US mine $/short ton	108.5	109.4	111.5	110.4	83.05	81.45
Real Dec 1991 prices	126.0	123.6	121.6	114.9	83.10	81.1
South African bulk fob (a) Rotterdam $/short ton	105-172	105-172	105-178	115-178	105-170	105-170
Raw bulk fob (a) US plant $/short ton	65-135.5	65-135.5	65-135.5	65-135.5	65-135.5	65-135.5

(a) Source: Industrial Minerals

South African material is generally higher quality than the US vermiculite.

Transport costs are important.

MARKETING ARRANGEMENTS

The Palabora Mining Company in South Africa and W R Grace & Co in the United States produce the larger part of the world's mined output. There are two other much smaller companies in the United States. Exfoliation plants are usually sited close to final markets for economic reasons.

REAL PRICES 1979 to 1991
Vermiculite, Average US value fob Mine

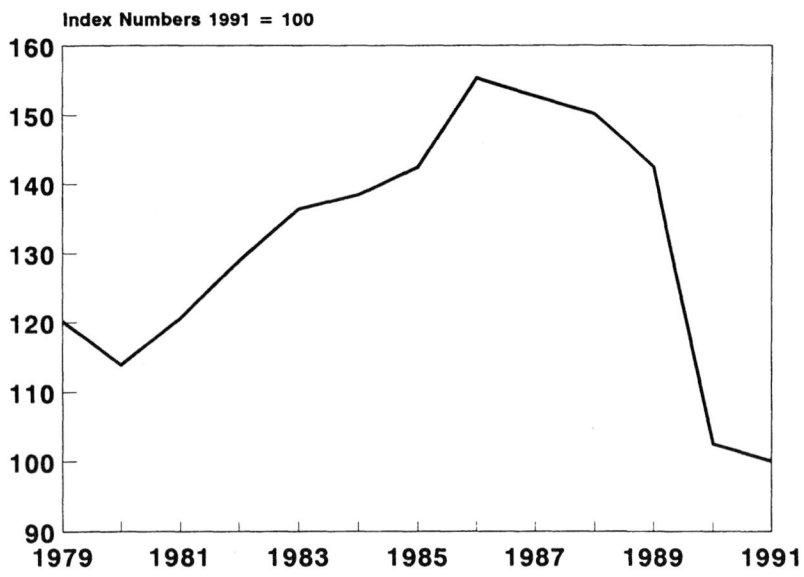

Index Numbers 1991 = 100

WORLD PRODUCTION 1979 to 1991
Vermiculite

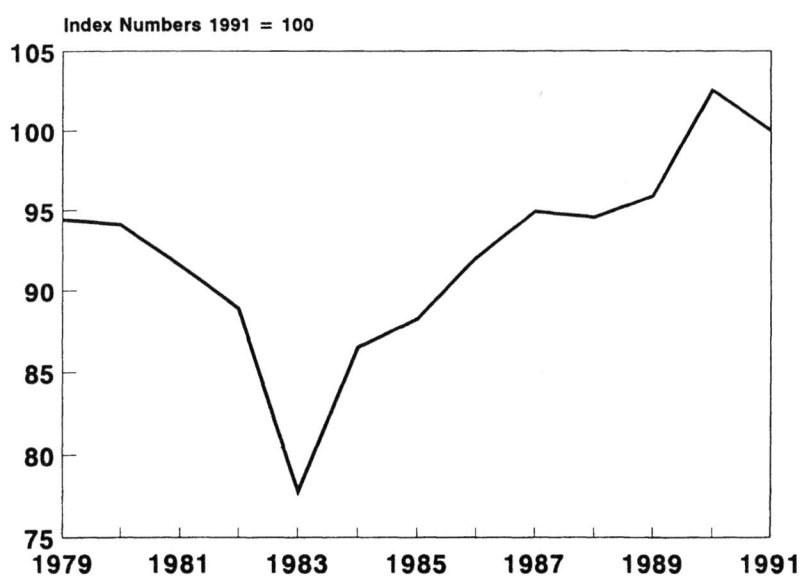

Index Numbers 1991 = 100

VERMICULITE

SUPPLY AND DEMAND BY MAIN MARKET AREA

	UK		EC(12)		Japan		USA	
	1989	1990	1989	1990	1989	1990	1989	1990
Production ('000 tonnes)	-	-	-	-	15	15	249.5	208.6
Net Imports ('000 tonnes)	45.5	52.1	265	291	33.3	37.0	50	45
Source of Net Imports (%)								
Hungary			6	6				
European Community	52	58					15	60
S Africa	31	29	36	36	56	53	78	34
China					42	45	6	
Turkey	15	10	31	32				
USSR			22	21				
Brazil			1					6
Others	2	3	4	5	2	2	1	
Net Exports ('000 tonnes)	1.4	1.1	69	95	21.3	24.6	18	18
Consumption ('000 tonnes)	c.44	c.51	c.200	c.200	c.27	c.27	281	236
Import Dependence Imports as % of consumption	100	100	100	100	n.a	n.a	18	19
Imports as % of consumption and net exports	100	100	100	100	n.a	n.a	17	18
Share of World Consumption (%) Total world	8	9	37	36	5	5	52	42
Consumption Growth (% p.a.)								
1970s	3.6		n.a		n.a		1.9	
1980s	n.a		n.a		n.a		-2.4	

ZINC

WORLD RESERVES
(million tonnes of contained zinc and % of total)

Developed			Developing			Centrally Planned		
Australia	18	(12.0)	Brazil	2	(1.3)	China	5	(3.3)
Canada	22	(14.7)	India	11	(7.3)	N Korea	4	(2.7)
Ireland	5	(3.3)	Iran	2	(1.3)	Poland	3	(2.0)
Japan	4	(2.7)	Mexico	6	(4.0)	USSR	10	(6.7)
S Africa	3	(2.0)	Peru	7	(4.7)			
Spain	5	(3.3)	Zaire	5	(3.3)			
USA	20	(13.3)	Other America	1	(0.7)			
Others	14	(9.3)	Other Africa	1	(0.7)			
			Other Asia	2	(1.3)			
Totals	**91**	**(60.7)**		**37**	**(24.6)**		**22**	**(14.7)**
Grand Total		**150**						

The reserve base totals 321 million tonnes and identified world resources total 1.8 billion tonnes. If hypothetical and subeconomic resources are included the total would be about 4.4 billion tonnes.

WORLD MINE PRODUCTION, 1989-90
('000 tonnes of contained zinc and % of total 1990)

Developed	1989	1990	% 1990	Developing	1989	1990	% 1990	Centrally Planned	1989	1990	% 1990
Australia	803.0	938.6	(12.9)	Argentina	43.2	38.7	(0.5)	Bulgaria	39.7	34.7	(0.5)
Canada	1216.1	1175.8	(16.1)	Bolivia	74.8	107.9	(1.5)	China	620.4	618.9	(8.5)
France	26.7	23.9	(0.3)	Brazil	105.8	110.0	(1.5)	N Korea	200.0	195.0	(2.7)
W Germany	63.9	58.1	(0.8)	Chile	18.4	25.1	(0.3)	Poland	170.0	154.8	(2.1)
Greenland	71.6	47.8	(0.7)	Honduras	33.7	29.3	(0.4)	Romania	11.0	11.5	(0.2)
Finland	58.4	51.7	(0.7)	India	65.2	70.0	(1.0)	USSR	940.0	870.0	(11.9)
Ireland	168.8	166.5	(2.3)	Iran	25.0	14.8	(0.2)	Czecho-			
Italy	44.0	42.4	(0.6)	Mexico	314.7	298.9	(4.1)	slovakia	6.6	7.4	(0.1)
Japan	131.8	127.3	(1.7)	Morocco	17.3	18.4	(0.3)	Vietnam	10.0	10.0	(0.1)
S Africa	77.3	74.5	(1.0)	Namibia	42.8	40.9	(0.6)				
Spain	265.3	257.5	(3.5)	Peru	598.1	585.1	(8.0)				
Sweden	168.0	159.9	(2.2)	S Korea	23.2	22.8	(0.3)				
Turkey	36.8	35.0	(0.5)	Thailand	86.6	80.8	(1.1)				
USA	288.3	538.2	(7.4)	Zaire	72.8	61.8	(0.8)				
Yugoslavia	75.0	76.0	(1.0)	Zambia	23.5	35.7	(0.5)				
Others	60.3	66.7	(0.9)	Others	21.6	16.3	(0.2)				
Totals	**3555.3**	**3839.9**	**(52.6)**		**1566.7**	**1556.5**	**(21.3)**		**1997.7**	**1902.3**	**(26.1)**
Grand Totals	1989-	7119.7									
	1990-	7298.7									

ZINC

WORLD SMELTER PRODUCTION, 1989-1990
('000 tonnes of zinc metal and % of total 1990)

Developed	1989	1990	% 1990	Developing	1989	1990	% 1990	Centrally Planned	1989	1990	% 1990
Australia	294.2	301.3	(4.2)	Algeria	28.0	23.6	(0.3)	Bulgaria	89.6	73.50	(1.0)
Austria	26.0	27.0	(0.4)	Argentina	31.6	31.5	(0.4)	China	450.9	526.3	(7.4)
Belgium	286.9	289.7	(4.1)	Brazil	162.3	154.1	(2.2)	E Germany	18.5	12.7	(0.2)
Canada	669.7	591.8	(8.3)	India	71.4	79.1	(1.1)	N Korea	259.0	239.0	(3.4)
Finland	162.5	174.9	(2.5)	S Korea	240.2	248.4	(3.5)	Poland	163.7	132.1	(1.8)
France	265.8	264.1	(3.7)	Mexico	194.5	199.0	(2.8)	Romania	29.8	11.5	(0.2)
W Germany	353.5	337.6	(4.7)	Peru	137.8	117.6	(1.7)	USSR	1020.0	920.0	(13.0)
Italy	246.0	248.1	(3.5)	Thailand	68.7	70.8	(1.0)	Vietnam	10.0	10.0	(0.1)
Japan	664.5	687.5	(9.7)	Zambia	12.9	10.8	(0.2)				
Netherlands	203.0	208.2	(2.9)	Zaire	54.0	38.2	(0.5)				
Norway	120.7	125.1	(1.8)								
Portugal	5.0	5.5	(0.1)								
S Africa	85.0	92.4	(1.3)								
Spain	246.4	252.7	(3.6)								
Turkey	24.2	20.1	(0.3)								
UK	79.8	93.3	(1.3)								
USA	358.2	365.7	(5.2)								
Yugoslavia	119.4	113.7	(1.6)								
Totals	4210.8	4198.7	(59.2)		1001.4	973.1	(13.7)		2041.5	1925.1	(27.1)
Grand Totals	1989-	7253.7									
	1990-	7096.9									

WORLD MINE AND SMELTER CAPACITY, 1990

('000 tonnes of metal)

	Mine	Metal
Developed		
Australia	1085	345
Belgium	-	345
Canada	1360	745
Finland	55	170
France	40	305
W Germany	70	410
Greenland	75	-
Ireland	170	-
Italy	76	253
Japan	160	864
Netherlands	-	205
Norway	20	130
S Africa	80	105
Spain	290	268
Sweden	180	-
UK	7	105
USA	605	360
Yugoslavia	90	160
Others	95	64
Total	**4458**	**4834**
Developing		
Brazil	170	169
India	75	99
S Korea	25	265
Mexico	320	218
Peru	625	172
Thailand	85	70
Zaire	80	72
Zambia	40	40
Others	462	85
Total	**1882**	**1190**
Centrally Planned		
North Korea	230	260
China	620	500
Poland	205	137
USSR	870	1165
Others	85	150
Total	**2010**	**2212**
TOTAL	**8350**	**8236**

In some cases effective capacities may be well below nominal levels.

ZINC

SECONDARY PRODUCTION:WESTERN WORLD
('000 tonnes of zinc 1989-90)

	1989	1990
Scrap used by primary smelters (and included in primary output)	403	419
Remelted zinc and alloys	205	169
Zinc in copper and other alloys	471	472
Scrap as such, used by chemical plants etc.	275	268
Total secondary recovery additional to smelter output	**951**	**909**
Total secondary recovery	**1354**	**1328**

Source:Metallgesellschaft.

RESERVE/PRODUCTION RATIOS

Static Reserve Life (years):	21
Ratio of identified reserve base to cumulative demand 1991-2010:	2 : 1

CONSUMPTION

	'000 tonnes		% p.a. growth rates		
	1989	1990	1960s	1970s	1980s
European Community	1585.4	1653.4	3.2	0.8	1.0
Japan	768.7	814.3	13.6	2.2	0.8
United States	1059.5	996.8	3.3	-2.4	1.3
Others	1776.9	1770.2	7.1	5.1	2.6
Total Western World	**5190.9**	**5234.7**	**5.1**	**1.4**	**1.5**
Total World	**7102.1**	**6979.1**	**5.1**	**2.3**	**1.0**

Data refer to consumption of slab zinc. They exclude remelted zinc.

END USE PATTERNS, 1990 (%)

	USA[1]	UK[2]	W. Germany[3]	Japan[4]
Galvanising	52	42	32	62
Die-casting alloys	20	18	21	14
Brass	13	21	23	13
Rolled zinc	-	1	17	-
Others	15	18	7	11

1. Slab zinc including secondary refined
2. Zinc in all forms including scrap
3. Refined zinc and direct use of scrap
4. Refined zinc only

Source: ILZSG

VALUE OF CONTAINED METAL IN ANNUAL PRODUCTION

$10.8 billion (as slab metal at average 1991 prices).

SUBSTITUTES

Aluminium, magnesium and plastics compete in some die-casting applications.

Ceramic and plastic coatings, electroplated cadmium and aluminium, and special steels compete in some galvanising applications.

Aluminium, magnesium and titanium can replace zinc in chemicals and pigments; zirconium is an alternative in ceramic and enamel applications.

Aluminium alloys, stainless steels and plastics can be used in place of brass.

TECHNICAL POSSIBILITIES

Development of new alloys, eg superplastic alloys of zinc and aluminium.

Increased recovery of secondary zinc.

PRICES

	1986	1987	1988	1989	1990	1991
US$/tonne						
LME Standard grade cash	706.9	-	-	-	-	-
LME High grade cash	754.4	798.2	1249.9	1655.0	1717.8	1519.3
European producer	798.8	825.3	1185.5	*	*	*
US producer	891.9	978.4	1359.0	1825.2	1662.7	1163.7
Real Dec 1991 prices						
LME Standard	-	-	-	-	-	-
LME High grade	875.9	900.6	1359.8	1720.2	1719.5	1513.2

*The European producer price was suspended in early 1989.

Most zinc metal outside is now traded at prices based on LME quotations following the demise of the European producer price in early 1989 and the switch of North American producers to LME based prices in 1991. Ores and concentrates are purchased by smelters at LME prices less negotiated treatment charges. Mine costs are influenced by by-product values.

ZINC

MARKETING ARRANGEMENTS

There is wide geographical spread of production with over 300 mines, but a much smaller number of smelters. Four or five large groups now dominate international production and trade yet there is a steady growth of smelter production near mines in developing countries, and new entrants to mining to counter, at least partially, this tendency towards increased concentration. The United Nations' International Lead and Zinc Study Group is an intergovernmental forum for statistical analysis and discussion of common problems.

REAL PRICES 1979 to 1991
Zinc, LME Standard/High Grade

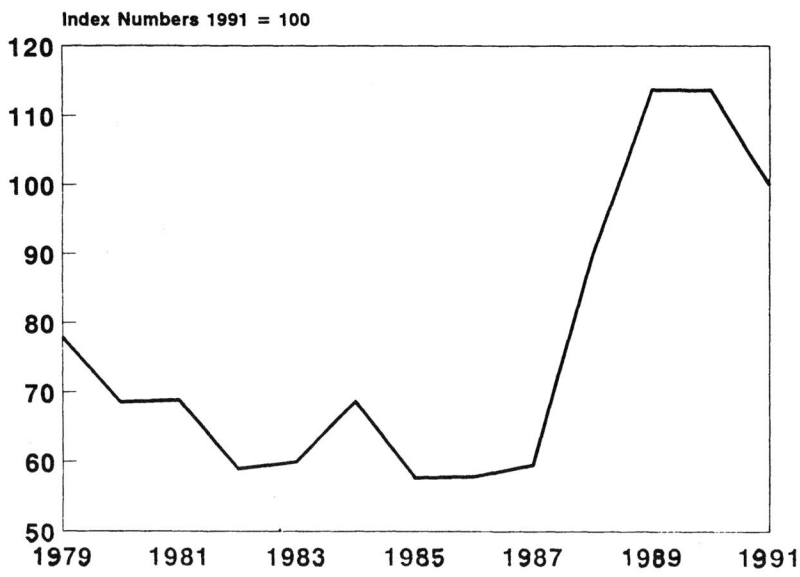

Index Numbers 1991 = 100

WORLD PRODUCTION 1979 to 1991
Zinc, Refined Metal

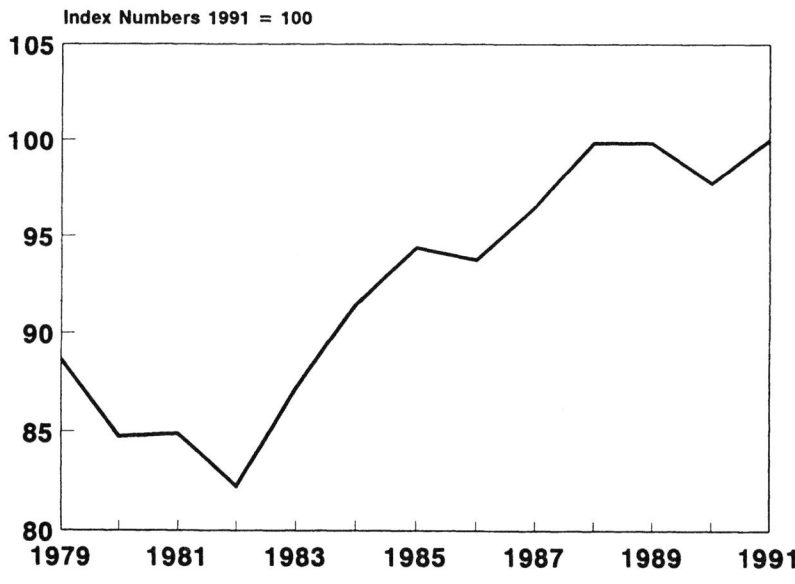

Index Numbers 1991 = 100

ZINC

SUPPLY AND DEMAND BY MAIN MARKET AREA

	UK		EC(12)		Japan		USA	
	1989	1990	1989	1990	1989	1990	1989	1990
Production								
('000 tonnes Zn content)								
Mine production	5.8	6.6	599.1	581.7	131.8	127.3	288.3	538.2
Smelter production	79.8	93.3	1686.4	1699.2	664.5	687.5	358.2	365.7
Zinc recovered from scrap (other than zinc in primary smelters)	49.6	52.4	518(a)	512(a)	176.3	161.3	249.3	245.6

(a) Assumes 60 for Belgium, Netherlands, Portugal and Spain combined.

	UK		EC(12)		Japan		USA	
Net Imports								
('000 tonnes Zn content)								
Ores and concentrates	79.2	86.5	1050.1	1214.2	523.1	586.3	388.6	51.8
Unwrought zinc metal	126.4	123.6	143.1	198.2	132.7	140.5	712.4	631.8

Source of Net Imports (%)

Ores and Concentrates	UK 1989	UK 1990	EC(12) 1989	EC(12) 1990	Japan 1989	Japan 1990	USA 1989	USA 1990
Australia	20	16	13	13	51	51	..	..
Canada	9	14	40	39	15	11	94	39
European Community	12	8						
S Africa (inc. Namibia)					1	1		
Sweden			8	6				
USA	9	10	4	5	3	7		
Bolivia			4	3	2	2		10
Honduras	20	7	2	2				
Mexico	7	1	6	7		1	4	47
Peru	21	33	14	15	15	19	1	4
Morocco		2	1	1				
China	1				8	3	..	
Poland		2		1				
Chile			3	3	2	2		
Others	1	7	5	5	3	3	1	

Metal	UK 1989	UK 1990	EC(12) 1989	EC(12) 1990	Japan 1989	Japan 1990	USA 1989	USA 1990
Australia		1			13	13	6	7
Canada	12	11	16	10	14	16	61	59
European Community	57	45			2		7	7
Finland	17	27	33	41			3	3
Japan								
Norway	13	16	36	31			4	4
Mexico					1	5	10	11
Peru					3	4	5	4
S Korea					26	18		
Zaire			1	3			2	1
Zambia			1					
China					8	10		
N Korea					28	32		
Poland				1				
Brazil					2		1	3
Others	1		13	14	3	2	1	1

	UK		EC(12)		Japan		USA	
	1989	**1990**	**1989**	**1990**	**1989**	**1990**	**1989**	**1990**
Net Exports								
('000 tonnes Zn content)								
Ores and concentrates	6.4	7.8	68.6	68.6	-	-	255.4	380.9
Unwrought zinc metal	8.9	7.7	167.9	144.1	17.9	21.3	5.7	2.3
Consumption								
('000 tonnes Zn content)								
Primary metal	194.5	189.0	1585.4	1653.4	768.7	814.3	1059.5	996.8
Zinc recovered from scrap	49.6	52.4	518	512	176.3	161.3	249.3	245.6
Direct use of concentrates	-	-	2	2	-	-	2.1	2.2
Import Dependence								
Imports as % of								
consumption	84	87	57	65	69	74	84	55
Imports as % of consumption								
and net exports	79	82	50	59	68	73	70	42
Share of World Consumption (%)								
(primary metal)								
Western world	4	4	30	32	15	16	20	19
Total world	3	3	22	24	11	12	15	14
Consumption Growth (% p.a.)								
1960s	0.7		3.2		13.6		3.3	
1970s	-3		0.8		2.2		-2.4	
1980s	0.4		1.0		0.8		1.3	

ZIRCONIUM

WORLD RESERVES
('000 tonnes of ZrO_2 and % of total)

Developed			Developing			Centrally Planned		
Australia	23300	(47.8)	Brazil	1000	(2.1)	USSR	4000	(8.2)
S Africa	13800	(28.3)	India	2100	(4.3)	China	500	(1.0)
USA	3200	(6.6)	Madagascar	125	(0.3)			
New Zealand	155	(0.3)	Sierra Leone	515	(1.1)			
Totals	**40455**	**(83.0)**		**3740**	**(7.8)**		**4500**	**(9.2)**
Grand Total	**48695**							

The reserve base is 58 million tonnes and includes, in addition to the above countries, deposits in Sri Lanka.

The world's identified resources exceed 60 million tonnes of contained zircon.

WORLD MINE PRODUCTION, 1989-90, and PRODUCTIVE CAPACITY, 1990
('000 tonnes)

	Mine Production Concentrates		% of total	Productive capacity
	1989	1990	1990	1990
Developed				
Australia	511	442	(50.0)	600
S Africa	180	180	(20.4)	180
USA	118	102	(11.5)	120
Total	**809**	**724**	**(81.9)**	**900**
Developing				
Brazil	33	33	(3.8)	35
India	17	18	(2.0)	20
Malaysia	19	4	(0.5)	20
Sri Lanka	3	3	(0.3)	3
Thailand	1	2	(0.2)	3
Total	**73**	**60**	**(6.8)**	**81**
Centrally Planned				
China	15	15	(1.7)	15
USSR	85	85	(9.6)	85
Total	**100**	**100**	**(11.3)**	**90**
TOTAL	**982**	**884**		**1071**

Concentrates typically contain at least 65% zircon (ZrO_2), and there is 74% of zircon in zirconium.

Some capacity has closed down or gone on standby since 1990.

RESERVE/PRODUCTION RATIOS

Static Reserve Life (years):	55
Ratio of identified reserve base to cumulative demand 1991-2010:	3.8 : 1

CONSUMPTION
(zircon concentrates)

	'000 tonnes		% p.a. growth rates	
	1989	1990	1970s	1980s
European Community	249	c.200	4.2	0.2
Japan	159	134	8.4	-3.4
United States	145	108	0.2	-2.5
Other Countries	275	c.250	n.a	6.0
Total Western World	**827**	**692**	**n.a**	**0.5**

Source: Industrial Minerals November 1990

END USE PATTERNS, 1990 (%)

Zircon	USA (1990)	Japan (1990)	Europe (1988)
Foundry sands	29	15	12
Refractories	24	52	16
Ceramics	n.a(a)	16	58
Other, including zirconia and abrasives, alloys, chemicals and metal for nuclear applications and chemical processing equipment	47(a)	17	14

(a) Ceramics included in other uses.

VALUE OF ANNUAL PRODUCTION

$272 million (at average 1991 prices).

ZIRCONIUM

SUBSTITUTES

Chromite, olivine and silica sand can be used in place of zircon in some foundry applications. Titanium and tin compounds can replace zirconium oxide in ceramics. Alumina, graphite and magnesia refractories compete with zirconium refractories. In Japan zircon refractories have lost out heavily to alumina.

A number of alternatives are available in the nuclear applications, notably stainless steel as a structural material and aluminium, niobium and vanadium for fuel containers.

Stainless steel, titanium and tantalum are substitutes in many corrosion-resistant industrial applications.

Many materials, particularly ferroalloys, compete with zirconium in ferrous metal applications.

TECHNICAL POSSIBILITIES

Replacement of zirconia abrasives by synthetic diamond and cubic boron nitride abrasives.

Potential use as a ceramic coating in aircraft engines and other applications where strength and high temperature oxidation are important. Development of zircon based ceramics for a wide range of uses.

PRICES

	1986	1987	1988	1989	1990	1991
Zircon sand 66-67% ZrO_2						
Australian Standard grade US$/tonne	108.6	150.9	245.8	412.8	461.1	307.7
Real Dec 1991 prices	126.1	170.5	277.8	429.1	462.6	306.6

MARKETING ARRANGEMENTS

Australia and S. Africa are the principal sources of concentrates. Metal production is in Japan, France and the USA. Zircon is mainly a by-product of beach sand production of titanium concentrates, so that its output is fairly inflexible in the face of changing demand.

REAL PRICES 1979 to 1991
Zirconium, Australian Zircon Sand

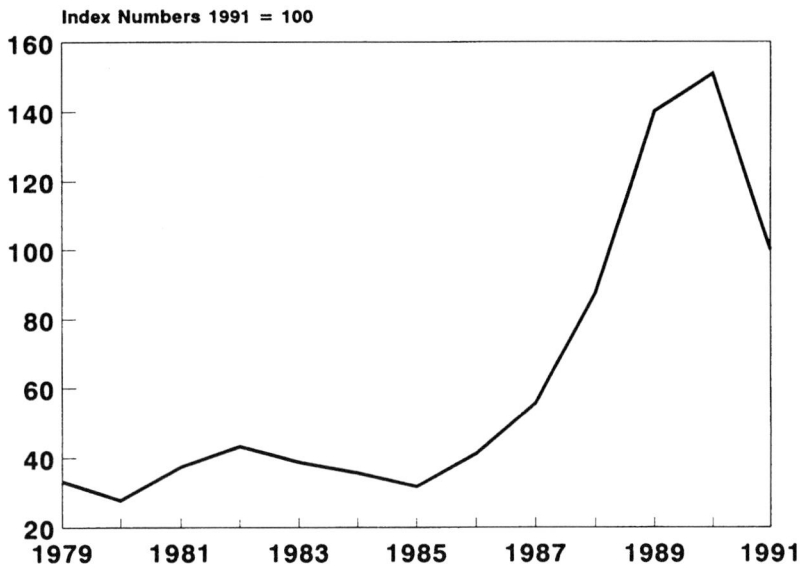

Index Numbers 1991 = 100

WORLD PRODUCTION 1979 to 1991
Zircon, Concentrates

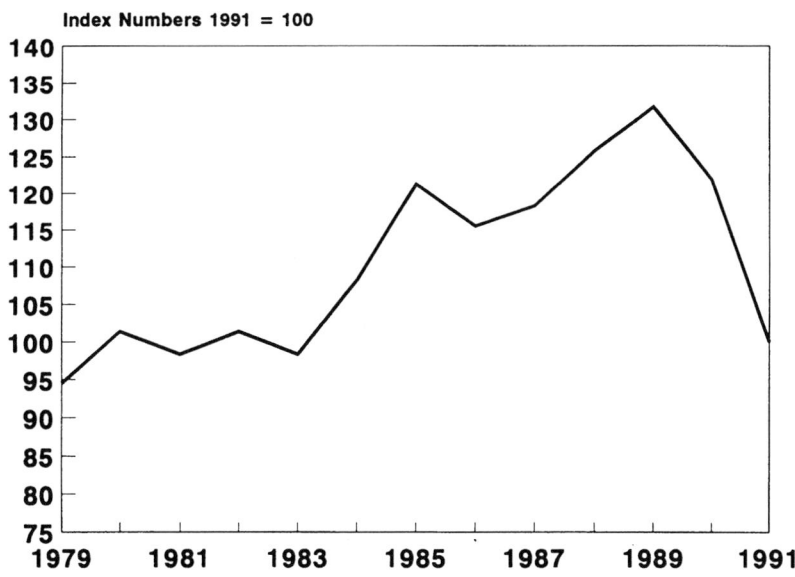

Index Numbers 1991 = 100

ZIRCONIUM

SUPPLY AND DEMAND BY MAIN MARKET AREA

	UK 1989	UK 1990	EC(12) 1989	EC(12) 1990	Japan 1989	Japan 1990	USA 1989	USA 1990
Production ('000 tonnes)								
Zircon concentrates	-	-	-	-	-	118	102	
Zirconium oxide	n.a	n.a	n.a	n.a	n.a	n.a	10.0	7.5 (excl. metal producers)
Net Imports ('000 tonnes)								
Zircon, ores & concentrates	55.2	35.7	264.8(a)	216.5(a)	172.8	119.0	73.1	28.3
Zirconium oxide	1.17	1.16	0.95	0.91	0.86(c)	0.99(c)	1.7(c)	1.8(c)
Zirconium sponge and metal	0.26	0.14	0.88	0.62	0.64	0.68	0.5	0.41

(a) Excl. Germany.
(b) Excl. Netherlands.
(c) Incl. germanium oxides.

Source of Net Imports (%)

	UK 1989	UK 1990	EC(12) 1989	EC(12) 1990	Japan 1989	Japan 1990	USA 1989	USA 1990
Ores and Concentrates								
Australia	55	22	61	62	81	73	70	42
European Community	2	11					1	1
S Africa	31	22	28	23	13	24	28	56
USA	1	1	4	3	2	2		
Sierra Leone	8							
Malaysia			2		2	1		
Brazil	1	1	2	1				
Other	2	43	3	7	2		1	1

	UK 1989	UK 1990	EC(12) 1989	EC(12) 1990	Japan 1989	Japan 1990	USA 1989	USA 1990
Net Exports ('000 tonnes)								
Zircon concentrates	0.8	0.9	13.54	18.2	0.4	0.3	50.6	30.2
Zirconium oxide	2.37	2.28	0.18(b)	0.10(b)	0.5(a)	0.7(a)	3.09(a)	1.78(a)
Zirconium metal	0.2	0.1	0.7(b)	0.4(b)	0.07	0.04	1.33	0.19

(a)Incl. germanium oxides.
(b) Excl. France.

	UK 1989	UK 1990	EC(12) 1989	EC(12) 1990	Japan 1989	Japan 1990	USA 1989	USA 1990
Consumption ('000 tonnes)								
Zircon concentrates	26	17	249	c.200	159	134	145	108
Zirconium oxide	n.a	n.a	n.a	n.a	n.a	n.a	8.6	8.5

	UK		EC(12)		Japan		USA	
	1989	1990	1989	1990	1989	1990	1989	1990
Import Dependence (based on zircon concentrates)								
Imports as % of consumption	100	100	100	100	100	100	50	26
Imports as % of consumption and net exports	100	100	100	100	100	100	37	21
Share of World Consumption (of concentrates) (%)								
Western World	3	2	30	29	19	19	18	16
Consumption Growth (% p.a.)								
1970s	-1.7		4.2		8.4		0.2	
(based on imports of concentrate)								
1980s	-6		0.2		-3.4		-2.5	

SOURCES AND NOTES

This handbook has drawn from a very wide range of primary and secondary statistical sources that are almost too numerous to mention. Frequently, different sources may give markedly different estimates for what is ostensibly the same figure, even when full allowance is made for differing definitions. In such instances judgement has been used. Widely varying units of measurement are used in the originals, and this handbook has standardised, with one or two exceptions on metric units.

The main sources were as follows:

World Reserves

For nearly all minerals the data are taken from US Bureau of Mines' sources, and mainly from:
Mineral Commodity Summaries 1992.
Mineral Facts and Problems. 1985 Edition. Bureau of Mines Bulletin 671
Minerals Yearbook Volume 1. Metals and Minerals 1989 and 1990 Preprints.

Production, Consumption and Trade

The same sources as for reserves, supplemented by:

World Mineral Statistics (1985-1989), British Geological Survey
United Kingdom Minerals Yearbook 1990 - British Geological Survey
Metal Statistics (1980-1990) - Metallgesellschaft, particularly for the main non-ferrous metals
World Metal Statistics (May 1992) - World Bureau of Metal Statistics
United Nations Monthly Bulletin of Statistics
Iron and Steel Statistics (1989 and 1990) - Eurostat
Lead and Zinc Statistics (May 1992) - International Lead and Zinc Study Group
UNCTAD Tin Statistics
Foreign Trade Analytical Tables for 1989 and 1990 (NIMEXE) - Eurostat
Statistics of Japanese Imports and Exports for 1989 and 1990
United Kingdom Trade Statistics for 1989 and 1990
United States Trade Statistics for 1989 and 1990
Yearbook of Mining, Non-Ferrous Metals and Products Statistics - MITI
Mining Year Handbook - MITI
Potassium and Phosphorus & Sulphur. Publications of British Sulphur Publishing
'Gold' - Gold Fields Mineral Services Ltd. (1992 edition)
Platinum Yearbooks - Johnson Matthey.
Silver - Handy & Harman (1991 edition)
Mining Annual Review 1989 to 1991 - Mining Journal Publications
Engineering and Mining Journal (March 1992 and April 1992 issues)
Australian Mineral Statistics (quarterly). Canberra
Tungsten Statistics - Annual Statistics of the UNCTAD Committee on Tungsten
TEX Report - Ferro-Alloy Manual
TEX Report - Iron Ore Manual
Aluminium, Copper, Lead, Nickel and Zinc World Flow Tables - World Bureau of Metal Statistics
Steel Statistics 1991. International Iron & Steel Institute
Uranium. Resources, Production and Demand (successive issues) - OECD Nuclear Energy Agency/IAEA
Uranium Institute. Uranium in the New World Market 1991 & 1992 Update.

Prices

Metals Week Handbooks 1982-1991
Metal Bulletin Handbooks 1982-1992
Industrial Minerals
Engineering and Mining Journal
USBM Mineral Yearbook and Mineral Commodity Summaries
IMF - International Financial Statistics
UN - Monthly Bulletin of Statistics
Sulphur - British Sulphur Corporation

Prices are yearly averages unless otherwise stated.

End Use Patterns

Minerals Yearbook 1990 Preprints
Mineral Commodity Summaries 1992.